BE YOUR OWN POWER COMPANY

Selling and Generating Electricity from Home and Small-Scale Systems

PHOTOVOLTAICS • WIND POWER
HYDROPOWER • COGENERATION

by David Morris
Illustrations by Frank Rohrbach

 Rodale Press, Emmaus, Pa.

About the Author

David Morris, president of the Institute for Local Self-Reliance in Washington, D.C., is also the author of *Neighborhood Power, Self-Reliant Cities, The New City States* and numerous magazine and newspaper articles. He has consulted with individuals, city governments and small businesses about the economics of small-scale power production.

The Institute for Local Self-Reliance is located at 1717 Eighteenth Street, NW, Washington, DC 20009. The Institute's phone number is (202) 232-4108.

Copyright © 1983 by David Morris

Printed in the United States of America on recycled paper, containing a high percentage of de-inked fiber.

Book design by Daniel M. Guest
Cover design by Daniel M. Guest
Technical special effects for the cover produced by
    Wernick Productions, Inc.
Cover photographs by Solarex Corporation (top), Mitchell T. Mandel
    (second and third) and Agway Research Center (bottom).

Library of Congress Cataloging in Publication Data

Morris, David J.
    Be your own power company.

    Bibliography: p.
    Includes index.
    1. Electric power-plants—Amateur's manuals.
I. Title.
TK9911.M63   1983   621.31'21   83-13821

ISBN   0-87857-477-8 hardcover
ISBN   0-87857-478-6 paperback

2 4 6 8 10 9 7 5 3 1 hardcover
2 4 6 8 10 9 7 5 3 1 paperback

CONTENTS

ACKNOWLEDGMENTS

First and foremost is my debt to Blue Mountain Center. Nestled in the heart of the Adirondacks, this 4,000-acre nature preserve offered me the opportunity to write this book in blissful isolation. To Harriet Barlow, the director of the center, I owe a more personal debt. Companion, colleague and friend, I can literally say this book would not have been written without her.

To John Plunkett I owe much of my understanding of the small power issue. As staff economist for the Institute for Local Self-Reliance, his assistance was immensely useful in helping me wade through the jargon and the statistical complexities.

My thanks go to the rest of the Institute staff, especially David Bardaglio who helped coordinate the project and keep all the pieces together.

Having written several books, I am constantly surprised by the varying time that publishers and editors take with authors. Rodale Press offered me not only time but an editorial oversight of extremely high quality as well. Through the painstaking efforts of Joe Carter and Margaret J. Balitas at Rodale, the manuscript was much improved. Editors at Rodale wear many hats. Having written his own book on solar technologies, Joe Carter's expertise proved very helpful to the final editing.

Of course, any errors or murkiness in this book are my sole responsibility. I hope the reader finds it a useful introduction to the new era.

INTRODUCTION

Electricity is no longer a luxury. It is essential to our well-being. A residence cut off from electricity in 1900 suffered the minor inconvenience of living by candlelight. Today cutting off one's electricity can be tantamount to murder. Witness the deaths of Dallas residents from lack of air conditioning during a recent heat wave or the deaths of Buffalo residents from lack of heat during a recent cold spell. We design for electricity and therefore are dependent on it even when it doesn't represent the primary supply source. For example, gas and oil furnaces are spurred into action by tiny amounts of electricity.

We are an electric society. Our news and entertainment come via electric-powered conduits. Our products are increasingly produced by electricity. Electricity is the foundation of our future computer and robotic society. Today almost one-third of all our primary energy (fossil fuels and uranium) is used to generate electricity.

Unfortunately we have almost no direct control over our electric system. Can you imagine depending solely on one company for your food supply, with no chance to grow your own or buy it from anyone else? Unthinkable. Yet until recently our access to electricity was limited in just this way.

In the early part of this century, utilities received the exclusive privilege of producing and selling electric power. Since then they have jealously guarded that prerogative, often disregarding certain responsibilities inherent in that privilege. While failing to meet the energy needs of rural America, the utilities vigorously opposed efforts to do so by such entities as the Tennessee Valley Authority (TVA) and Rural Electric Cooperatives (REC). It is a matter of record that an act of Congress, not private enterprise, finally brought low-cost and reliable electric power to American farms.

In this same period, when urban dwellers tried collectively to own their own electric systems, the utility response was equally negative. A national campaign that was unwittingly funded by utility customers smeared the municipal utility movement as communist-inspired. Bills were introduced in state legislatures to make municipal ownership illegal.

When cities were successful in their efforts, they soon found that while ownership of the distribution lines increased their leverage, it wasn't enough. The larger utilities still controlled the power plants and manipulated prices at will. If a city-owned utility tried to get a better price from another producer, the investor-owned utility invariably refused permission to transmit the cheaper electricity over its lines. And woe be it to any individual or enterprise that wanted to produce power independently. Utilities cut them off or charged them exorbitant prices for back-up power.

The vigor with which utilities protect their monopoly hasn't waned, but the environment within which they try to wield that power has indeed changed — dramatically. Technological developments and changes in the cost of conventional electric generation in the 1970s laid the foundation for a profound transformation of the electric generation system. The winds of change are blowing through the utility structure.

Today conservation is cheaper than production. Every time the utility raises its prices, its customers reduce their demand. Almost half a trillion dollars in projected power plants have been canceled in the last ten years. With aggressive conservation investments, utilities may well never need to build another large central power plant.

We can never conserve to the point where we use no electricity. New power plants will be necessary to replace old ones. But we would be better off building thousands, even millions, of small power plants. The rising costs of conventional power plants and an increasing maturity of renewable-based power plants and small cogeneration systems make this approach economically feasible. Evidence exists that many small plants rather than a single mammoth facility can increase the whole system's overall reliability. We are in a time of much change, when we cannot accurately predict future events. Having a great variety of small power plants that can be quickly brought on-line gives us a greater degree of flexibility and responsiveness to external changes such as in energy demand and pricing.

The economic attractiveness of decentralization is becoming ever more apparent. Yet to emphasize only the economic value of decentralization would be a mistake. The political and psychological value of a widely distributed capacity to produce a commodity as essential as electricity is equally important. Self-reliance was a major objective of the nation's founders. Benjamin Franklin once remarked, "The man who trades independence for security usually deserves to end up with neither." In the beginning of this century we voluntarily relinquished control over electric production in return for a lower cost and more reliable supply from utilities. Today that decision has brought us not only dependence, but insecurity as well.

In 1978 in an act of unprecedented political foresight, the U.S. Congress gave the American people the opportunity to reconstruct our electrical system with an emphasis on local self-reliance. The Public Utility Regulatory Policies Act (PURPA) abolished the monopoly utilities have over electric generation. It forced utilities to purchase electricity from independent power producers at a fair price.

That act presented the opportunity to decentralize control over electric generation. It did not make such decentralization inevitable. Congress did not enact PURPA out of a desire to promote small power plants. PURPA was designed to minimize the use of oil in electric generation by encouraging high-efficiency power production and the use of renewable fuels. Congress encouraged the growth of independent power producers only because it had become convinced that utilties would not promote the new technologies rapidly enough. However, small-scale systems and self-reliance were not objectives recognized by Congress.

PURPA makes possible a self-reliant future but does not ensure one. For instance, one of the incentives PURPA offers the independent producer is exemption from state and federal utility regulations. That incentive presents a danger as well as an opportunity for those desiring more control over the electric system. Currently the vast majority of investment in PURPA-sponsored power plants is for relatively large facilities owned by major corporations. Conceivably by 1995 we could be just as dependent on outside institutions for our electricity but the producers will not be regulated at all.

PURPA abolishes the monopoly utilities have over electric power production, although their monopoly over the distribution systems continues. PURPA doesn't allow independent producers to send electricity along the utility lines to another buyer. When that next step is taken and utilities are required to transmit electricity from one producer to another buyer, the grid system will become a gigantic marketplace. The price will be established through the actions of hundreds of thousands of buyers and sellers of electricity.

To transform the grid system into a marketplace and to add the ethical considerations of small-scale systems and self-reliance to PURPA requires a strong political movement. Such a movement is quite possible. Each household power plant produces very little electricity. But each household contains two voters. A hundred thousand such plants may generate less electricity than one central power plant.

But they represent an extremely powerful political constituency if organized, a constituency to demand the best price from the utility and to become actively involved in restructuring the electric system around the small power producer.

Many of those attracted by the prospect of owning their own power plants have a disdain for politics, preferring to work alone rather than in groups. Others view negotiations as a scientific process. The utility presents its data. The small power producers present their data. And truth emerges.

But the fact is that without collective effort, the small power movement can make little overall impact because it cannot change the rules of the game. The negotiation process is not an objective pursuit of ultimate truth but a bargaining process between two parties with very different objectives and visions of the future.

I remember finishing the chapter of this book titled "Interconnecting with the Grid" while sitting in the magnificent dining room of the Blue Mountain Center, a writer's colony in the heart of the Adirondacks. A typical August downpour was raging outside, the rolling thunder punctuated by jagged bolts of lightning. I was reading a contract between the local utility and the owner of a small wind turbine. Worried that the "dirty" electricity coming from the wind machine would contaminate the grid system, the utility demanded guarantees of purity. Three pages of the contract described the standards for the small producer. One paragraph caught my attention. The wind machine owner had to provide an oscillographic print of the waveform coming from the machine.

That the utility wanted to protect its customers from low-quality electricity was laudable. But the situation in which I found myself reading these lines indicated the utility clearly didn't apply those rules to itself. For three hours lights had flickered and my word processor had gone crazy. An hour before, the system had collapsed. I was reading the contractual provisions on waveform purity by the light of hurricane lamps.

Stories of the double standard are legion. Representatives of one large utility appeared before the California Public Utility Commission in two unrelated cases. In one they were arguing in their capacity as sole seller of electricity for a rate increase. To justify the increase the representatives presented data that showed that the next hydroelectric plant the utility was to build would cost more than 15¢ per kilowatt-hour produced. In the other case, the same utility was arguing (in its capacity as the sole buyer of electricity from small power plants) that they should be paying an independently owned hydroelectric facility owner only 4¢ per kilowatt-hour.

This book is intended as a technical aid to those desiring to enter the power production business through the doors opened by PURPA. It translates jargon, examines the economics of the utility business, discusses legal issues and describes the state of the art for small-scale electric plants. Knowledge is a necessary precondition to entering this business. But—I say again—without a collective effort by all those desiring to regain some control of the electric system, this technical knowledge will prove insufficient.

We are in the throes of a revolution in the way we generate and transmit electricity. The only certain prophecy is change. The electric utility in the year 2000 will be organized differently and play a different role. But as Bertrand Russell once remarked, "Change is one thing. Progress is another. Change is scientific; progress is ethical. Change is indubitable, whereas progress is a matter of controversy." Will we have change or progress? The answer rests in the strength of the grass-roots movement to harness new energy technologies to bring under popular control a resource that is basic to our well-being.

CHAPTER 1
The Electric Revolution

The age of electric power began in 1800 when Alessandro Volta, a professor of natural history at the University of Pavia in Italy, announced the discovery of a new form of electricity in a paper entitled "On the Electricity Excited by the Mere Content of Condensing Substances of Different Kinds." Before Volta the only kind of electricity known was *static electricity*—the kind you get when you rub your shoes across a carpet and then touch a conductor. In fact, the word *electricity* comes from the Greek word for amber, *elektron*. The Greek philosopher Thales discovered that amber rubbed with a cloth has the power to attract light bodies such as feathers, leaves, straw and small bits of wood.

Static electricity was even harnessed to perform useful work. More than a hundred years before Volta, static electricity was used in the first electric machine. A sulfur ball turned by a crank on an axis was excited by the friction of the hand and produced the first electric light.

Volta's remarkable contribution to the development of modern electricity lay in his discovery of "current" electricity. He transformed electricity from a toy to a tool of vast potential. By chemical means he produced a steady current. He created the first electric battery by alternating disks of silver and zinc piled one on the other. Each pair was separated from the adjoining pair by a cloth or paper disk saturated in brine. From the ends of this pile Volta could draw a continuously flowing electric current.

The "voltaic pile" unleashed a wave of discoveries and inventions throughout the world. Experimenters and scientists quickly refined Volta's crude invention into tiny power plants with which to conduct experiments. Gradually they began to uncover the basic theoretical principles underlying electric power and to conceive the mathematical equations that coupled electricity and magnetism. Technicians designed devices to harness this new source of energy. Electric current was used to decompose water, to cause charcoal to glow with an intense light, and later to deposit metals by electrolysis.

Hans Christian Oersted discovered in 1820 that a wire connected to the ends of a voltaic pile was enveloped by a magnetic field. If the wire was looped into a coil, the magnetic strength of the field was greatly intensified. André Marie Ampère proposed that the effect could be used to transmit messages over great distances. By 1836, practical systems of electric telegraphy were developed by Wilhelm Eduard Weber and Karl Friedrich Gauss in Germany, Sir William Fothergill Cooke and Sir Charles Wheatstone in England, and Joseph Henry and Samuel Morse in the United States.

Soon after Oersted's discovery of the magnetic field of currents, Michael Faraday, the great English chemist and physicist, began to investigate the subject. In 1831 he found that when a current is started in a coil of wire, a momentary current is induced in another nearby coil. When the primary current is stopped, an induced current is again generated, but in the opposite direction. He demonstrated that the effect is due to the magnetic field of the primary current and that the induced current in any circuit is proportional to the rate of change of the number of lines of magnetic force cutting through the circuit.

Producing magnetism from electricity opened the way to converting mechanical energy into electrical energy. By wrapping a core in coils of wire and turning the core through stationary magnetic fields, an electric current could be induced. Thus, the *electric generator,* or dynamo, and its important adjunct, the *transformer,* were born.

That same year, Joseph Henry, a physics teacher in the Albany Academy in New York, constructed the first electromagnetic motor. He increased the lifting power of a magnet from 9 to 3,500 pounds. Indeed, every electric dynamo and motor now uses the electromagnet in virtually the same way that Henry's motor did.

A year later, Thomas Davenport, an inventor in Brandon, Vermont, perfected the first commercially successful electric motor. It weighed 50 pounds and turned at 450 revolutions a minute (RPM), cutting through the magnetic fields with each revolution. The same year Hippolyte Pixii developed the first practical generator. By coupling a steam-driven turbine to the generator he could produce electrical energy, freeing the electrical experimenter from a reliance on chemical batteries.

These breakthroughs commercialized electricity. Motive power could be sent long distances. To understand what this means, consider that in 1851 a 1-inch-diameter shaft could transmit perhaps 1 horsepower (HP), or 0.75 kilowatts (kw), with a bearing every 3 feet. In less than a mile, all that power would have been consumed in bearing friction, even with the finest bearings then available. By comparison, a 1-inch-diameter shaft made from copper could conduct over 2,000 amperes of electricity. At 115 volts (V), this copper wire could conduct

363 HP for a mile with very modest energy losses. Since there was no torque, no bearings were needed. Moreover, the conductor could be suspended from widespread poles or towers.[1] Is it any wonder that some people envisioned electricity as the tool that would make humans titans?

Industry was quick to pick up on the potential of electric power. Its first major commercial use was for lighting. The *arc light* was developed in 1810. The device forced an electric voltage to leap across a gap between two wire tips, producing a brilliant arc of light 5 inches long. By the 1860s, steam-driven generators lit arc lamps in Europe and the United States.

But the arc light had several key drawbacks. The tips burned away in less than a night, and the brilliant, glaring light was suitable only for illuminating streets or very large indoor spaces, such as theaters or factories. Even more restricting was the practice of linking each light in series. If one bulb burned out, the whole system went dark.

Edison and the Rise of the Modern Utility

Enter the first electric entrepreneur—Thomas Alva Edison. Edison was a pragmatic inventor. Fresh from his triumphant innovations with the telegraph and the phonograph, he knew that the first step was to broaden the market for electricity. He knew immediately that the stumbling block was the electric industry's inability to bring electricity into individual buildings. The central problem was the use of series wiring. Edison later wrote, "I saw what had been done had never been made useful. The intense light had not been subdivided so that it could be brought into private homes."[2]

After only two nights of intense experimentation, he hit on a solution. By designing circuits in parallel, in effect duplicating the paths that electrons could flow through, he allowed the system to continue to function even if one individual component went dead.

Having solved the problem of subdividing current, he proceeded to refine the first electric consumer product—the light bulb. After thousands of tries that gave credence to his motto that invention is 1 percent inspiration and 99 percent perspiration, Edison developed an incandescent light bulb with a very fine filament of carbon inside an evacuated bulb. The high resistance in the wire generated heat and light when a relatively low current was passed through it. The filament lasted much longer than the tips of the arc lamp.

Four years after he began his search for a better light bulb, Thomas Edison unveiled the first central electric power station, in downtown

Photo 1–1:   Edison, shown here in his laboratory, had an important role in the creation of the modern electric utility. *Photograph courtesy of the Edison National Historic Site.*

series wiring (voltage is divided)

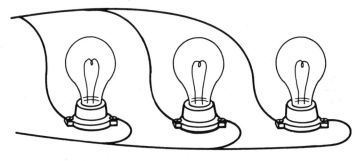

parallel wiring (voltage is constant across all bulbs)

Figure 1–1: If light bulbs are wired in series, the failure of any one bulb breaks the entire circuit. Parallel wiring, which Edison called *subdividing*, allows a continued flow of current even when one bulb fails.

Manhattan. The Pearl Street Station served 12 city blocks and was powered by steam generated from coal. Its initial capacity was 900 kw.

But steam was not the only force that could be used to drive a turbine. Commercial power plants immediately tapped into the kinetic energy of moving water. Mills that previously used on-site waterwheels to generate mechanical power now installed electric power plants. The first commercial hydroelectric plant was established in Appleton, Wisconsin, a year after the Pearl Street plant. Its original waterwheel measured 42 inches in diameter. It operated under a 10-foot head and had a speed of 72 RPM. Two Edison K dynamos were used, each capable of powering 250-candlepower of lighting, equivalent to a rating of 12.5 kw.

These early plants had none of the refinements of their modern successors. The Appleton plant had no voltage regulators. Operators depended on their eyes to gauge the brightness of the lamps. There were no meters and no fuse protection. Customers were charged by the lamp regardless of the hours of use. The original customers paid about 33¢ per lamp per month for service that lasted from dusk to dawn. Bare

Figure 1–2: Pearl Street Station in downtown Manhattan was the nation's first central power plant. *Reprinted from* Scientific American, *26 August 1882*.

copper wire was used in the distribution lines. Needless to say, there was no uniform safety code to regulate the proper use of electric power.

Despite all his inventiveness, Edison did not have the electric power field all to himself. Strong competition arose. In its infancy, the industry sold complete systems in which every part was patented, from the bulbs to the power plant components to the relays and switches. Equipment purchased from one supplier wasn't always compatible with another's electrical system. Different motors operated on different frequencies.

But even with the handicaps caused by a lack of standardization, electric power immediately captured America's fancy. By 1890 a thou-

This room is equipped with Edison Electric Light. Do Not Attempt to Light with Match. Simply Turn Key on Wall By the Door. The Use of Electricity for Lighting is in No Way Harmful to Health, nor Does it Affect the Soundness of Sleep.

Message placed wherever electric light was present in the 1870s and early 1880s.

sand central stations were operating. Department stores, local governments and industries were lighted with electricity.

But it was in the transportation sector that electricity found its greatest use. The first trolley system began in 1888. Two years later, 51 municipalities had electric streetcars. By 1895 electric trolleys operated in 850 cities on more than 10,000 miles of track. The electric streetcar companies remained electricity's biggest customer until 1920.

The streetcar was well suited to Edison's power plants because they generated *direct current* (DC), that is, current moving in one direction only. Direct current's chief disadvantage was that its voltage couldn't be easily raised or lowered. Therefore the voltage that left the power plant had to be used by the customer. But commercial customers used relatively low voltages (110 v to 220 v), and low voltages could be sent only a short distance because of energy losses related to low-voltage transmission. Streetcar companies, however, used high voltages, and they tended to own local power plants. Until the electric industry could learn how to raise and lower voltages, electric power sales would depend on many small, dispersed power plants. Edison's utility systems could not economically transmit electricity more than 2 miles.

Soon after Edison began building DC power plants, George Westinghouse began experimenting with another form of electricity, *alternating current* (AC). Alternating current, with its alternating negative and positive voltages, moves back and forth, and its voltage can be raised (stepped up) or lowered (stepped down) by using transformers. By changing the ratios of the windings in the primary and secondary coils of a transformer, the voltage could be changed in any direction. In 1885 Westinghouse purchased the English patents to a series AC distribution system developed by Lucien Gaulard and John D. Gibbs. While working for Westinghouse, William Stanley improved the Gaulard-Gibbs system by designing induction coils (later called transformers) in parallel connection and developed the AC, constant-potential (voltage) generator.

In March of 1886 Stanley demonstrated the practicability of alternating current in Great Barrington, Massachusetts, by transmitting single-phase electric power a distance of 4,000 feet using one transformer to increase the output voltage to 3,000 v and another to reduce it at the receiving end to 500 v. Later in 1886, Westinghouse installed in Buffalo, New York, the first commercial AC system.

With Westinghouse's advance in generating and distributing alternating current, the supply side of the modern electric utility was in place. Within a year, between 30 and 40 plants were in successful operation. However, to make full use of alternating current, what the modern electric system now needed was a practical motor that would run on alternating current. It was developed through the genius of Nikola Tesla, a Serbian immigrant who had once worked with Edison. He developed the first AC motor and may be viewed as the true father of our modern electric system. As one admiring biographer noted, "He conceived of such practical alternating-current motors as polyphase induction, split-phase induction and polyphase synchronous as well as the whole polyphase and single-phase motor system for generating, transmitting and utilizing electric current. And indeed, practically all electricity in the world in time would be generated, transmitted, distributed and turned into mechanical power by means of the Tesla Polyphase System.[3]

Tesla's lecture on AC motors on 16 May 1888 before the American Institute of Electrical Engineers was a landmark event. With one lecture Tesla had literally set the stage for the new era of electric power generation that would utilize AC power. Tesla's motor used the principle of the rotating magnetic field produced by two or more alternating currents out of step (out of phase) with each other. By creating, in effect, a magnetic whirlwind produced by the out-of-step currents, he eliminated both the need for a *commutator* (the device used for reversing the direction of an electric current) and for brushes that provide for the passage of the current.

Westinghouse immediately purchased the patents to Tesla's motor and his polyphase system. Tesla's motors operated on 60-cycle current, so Westinghouse altered his entire generation system from 133 cycles to 60 cycles to accommodate Tesla's design. Today 60-cycle current is still the standard in the United States.

The battle between Edison and Westinghouse was fierce but short-lived. Edison initially called alternating current "the killer current" and waged a great but unsuccessful publicity campaign against its introduction. The only remnant from that struggle exists in New York, where in the early 1890s prison authorities agreed to adopt the electric chair, which used alternating current. Edison used to say that anyone dying in the electric chair had been "Westinghoused." Edison left the electric power plant business shortly after his company merged with another to form General Electric. Today General Electric and Westinghouse remain the two major suppliers of electric equipment in the United States. Edison may not have been the father of the modern electric utility but, as always, he foresaw the development of companies that sold electricity rather than power plants. In 1883 he patented the first electric meter. Five years later, O. B. Shallenberger invented the ampere-hour meter for measuring alternating current.

Alternating current opened the way for higher voltages and longer transmission lines. Remote waterfalls and rivers became large generators of electricity. In 1896 three Westinghouse 5,000-horsepower (HP) turbines rotated by the force of Niagara Falls sent some 12,000 kw of power surging across lines built by General Electric to run lights, streetcars and motors in Buffalo, 26 miles away.

From its inception the electric generation and distribution industry was widely viewed as the key to economic growth. It was also an extremely profitable business. During its first 30 years, the relationship between the electric industry and government was constantly changing, driven by the changing nature of the technologies underlying the industry. Two key issues formed the focus of public debate. Would the industry be a monopoly or would it be competitive? And, who would own and regulate the industry?

The first electric utilities were small, neighborhood businesses. The industry was private and largely unregulated. Typical of this competitive period was the granting by the Denver Common Council in 1880 of an electricity franchise "to all comers" with the sole restriction that these companies not block public streets and roads.

Cities often granted multiple franchises. Chicago, for example, had more than 29 electric utilities operating within its boundaries in the late nineteenth century. New York City awarded 6 franchises in a single day in 1887.

The modern electric utility operating under what is essentially a monopoly awarded by a city or state is a creature of the new steam turbine technology introduced at the end of the nineteenth century, which, because of its increasing scale, made larger plants feasible. The first steam turbine, a 2,000-kw plant, was installed by Westinghouse for the Hartford Electric Company in 1900, revolutionizing the generation of electricity from coal. In 1903 the Commonwealth Electric Company installed a 5,000-kw plant. Eighteen months later the country's largest power plant generated 10,000 kw, and by 1913, a 35,000-kw plant was operating. In the mid-1920s, a single power plant could generate 175,000 kw, enough to meet the needs of a small city.

High voltage lines carrying alternating current permitted long-distance power distribution. In 1907 E. M. Hewlett and H. W. Buck developed the first suspension insulators, making practical the transmission of very high voltages. By 1920 voltages up to 132,000 v, or 132 kilovolts (kv), were common, and some lines operated at 150 kv. By 1934 the Hoover Dam transmitted 287 kv to Los Angeles, a distance of 270 miles.

Samuel Insull, a former secretary and salesman for Thomas Edison, is considered the father of the modern utility. As president of Commonwealth Electric Company he justified a monopoly on the basis of technological advances. When he took office in 1900, almost two-thirds of the nation's electricity was generated on-site, primarily by streetcar companies and other commercial and industrial producers. His goal was to consolidate all the small electric utilities into one big company and to persuade those who owned their own power plants to abandon them and buy cheaper power from the emerging grid system.

Insull pointed to the higher efficiencies of the newer, larger steam turbines. He also argued that electric demand had an important "diversity factor." People tended to use electricity at different times. Therefore, he said, the increase in the number of users was not proportional to the increase in generating capacity. Insull's favorite illustration was of a block of northside Chicago homes. There were 193 apartments on that block, and 189 of them were customers of his utility. There were no appliances, motors or other electrical devices to speak of in that block of apartments—just electric lamps. The power demanded by all separate apartments on the block, if totaled, was 68.5 kw, but since different lamps would be in use at different times, the actual maximum demand was only 20 kw.[4]

Supplying all of those customers from a single source would thus require only a 20-kw peak generating capacity. But if each household were equipped with a separate generating plant to meet its own needs, 68.5 kw would be needed—more than three times as much.

Insull backed up his rhetoric with an attractive pricing structure for large customers. His was the first *declining block rate*—the more you used, the lower the price per kilowatt-hour. In 1915 Chicago's residential customers paid 15¢ per kilowatt-hour. Its industrial off-peak customers paid only a penny.

Insull's persuasive sales pitch, combined with promotional pricing and the increased efficiencies of steam turbines, proved an unbeatable combination. In 1900, 60 percent of electricity was generated on-site, but by 1920 only one out of five kilowatt-hours (kwh) of electricity was generated on-site. From 1919 to 1927, 52,000 steam engines were scrapped; 18,000 internal combustion engines were discarded; and 5,000 water wheels were abandoned. Plugging into the utility monopoly had become cheaper than producing power on-site.

By the early twentieth century, the issue of whether electric power would be produced and distributed through monopoly or competition was decided. It would be a monopoly. No distinction was made between a monopoly of the transmission and distribution system and a monopoly of the power generation system.

The organizational form was clear. Less clear was the answer to the second question. Who would own and control the electric monopoly?

Smaller cities and rural areas had fewer potential customers, which made these markets less profitable for investors. Smaller cities were forced to finance and build their own power plants to satisfy a growing demand. In 1896 there were 400 municipally owned electric plants, and by 1906 there were more than 1,250. At the same time, about 2,800 investor-owned utilities accounted for slightly less than 75 percent of the generation capacity of the country.

While smaller cities were becoming directly involved in the generation and distribution of electricity, the larger cities, where private ownership predominated, were relinquishing direct oversight responsibilities. The issue of public ownership in the larger cities was central to most municipal elections during the early part of this century. Elections were won or lost on one's stance toward the electric utilities (and their direct brethren, the electric transportation or traction companies). In most cities municipal ownership movements eventually failed, although in several, such as Los Angeles, Seattle and Cleveland, they were successful.

Cities then as now retained the right to allow an electric company to operate within their jurisdictions by issuing a franchise to sell electricity within their borders. The awarding of franchises was among the most corrupt events in local politics. Many franchises were voted "in perpetuity." Later courts and state legislatures overturned these permanent franchises, but in several states 50- to 90-year franchises are not uncommon even today.

City councils at first also directly regulated the utility, setting rates and operating conditions. But as utility industries grew more complex and the technology permitted regional and even interstate distribution systems, the need for greater expertise and the lack of system-wide control undermined municipal authority. Moreover, the political process of oversight often culminated in political corruption and drawn-out court cases. A national movement arose to have independent state agencies regulate the utilities. It was led, ironically, by Samuel Insull, who consistently preached to his associates that only by allowing independent regulation could the industry hope to have the public accept its monopoly status.

The movement toward independent state regulatory commissions was fought by those involved in the municipal home-rule movement. A coalition of urban residents fighting for greater political autonomy from their state legislatures also fought for direct control over their physical infrastructure, their water systems, energy systems, roadways and transportation systems.

Those who argued for removing regulation to an independent state authority emphasized the efficiency of such a move. Those who supported regulation by the cities—such as Stiles P. Jones, a utility expert with the National Municipal League—considered democratic government, not scientific regulation, to be the goal. To him:

> *"Efficiency gained at the expense of citizenship is a dear purchase. Efficiency is a fine thing but successful self-government is better. Democratic government in a free city by an intelligent and disinterested citizenship is the greater ideal to work to and democracy plus efficiency is not unattainable."* [5]

But the dynamics of the technology undermined the arguments of even the most ardent supporters of local regulation. Even the most fervent believers in municipal home rule, such as Delos Wilcox, author of the two-volume text for citizen activists entitled *Municipal Franchises,* finally conceded that "public utilities, although still comparatively simple industries, have grown far enough beyond merely local bounds to require complex governmental machinery to operate or regulate them." [6]

Municipalities continued to own utilities, as later did rural electric cooperatives. The relationship between these publicly owned utilities and state regulatory bodies was, and continues to be, inconsistent, differing state-by-state.

Cooperative utilities are regulated in about two-thirds of the states. Municipally owned utilities are subject to the general jurisdiction of regulatory commissions in only nine states (Maine, Maryland, Nebraska, New York, Oregon, Rhode Island, Vermont, West Virginia and Wisconsin). Some states have unique statutes. For example, until 1981 a city in Illinois could regulate the local operation of public utilities if the electorate chose to do so through a referendum. In Kansas, local governments can still regulate public utilities that operate within a single municipality.

This inconsistency became important in the 1980s when small power producers found that, in some states, one state regulatory body would set prices for independently produced power, while in others, many individual publicly owned utilities retained that authority.

As transmission voltages increased and transmission lines fanned out, even the authority of state regulatory commissions was undermined. By 1935, 20 percent of the nation's electricity crossed state lines.

Such distribution systems made electricity part of interstate commerce and thus immune from state regulation, according to the Constitution of the United States. This was made clear by the U.S. Supreme Court in a 1927 case. The Narragansett Electric Lighting Company of Rhode Island sold a small amount of electric energy to the Attleboro Steam and Electric Company of Massachusetts. Because the Rhode Island Commission believed that the selling price was so low as to put an unjust burden on its other Rhode Island customers, it sought to raise the rate to the Massachusetts wholesale customer. But the Supreme Court held that the order of the Rhode Island Commission raising this specific rate constituted an unconstitutional burden on interstate commerce.

Into this regulatory vacuum stepped the Federal Power Commission, established in 1920 under the Federal Water Power Act. The Federal Power Act of 1935 consisted of amendments to the 1920 legislation, expanding the jurisdiction of the commission by giving it power to regulate the rates and service of electric utilities when the transactions are in interstate commerce. In the late 1970s the Federal Power Commission was renamed the Federal Energy Regulatory Commission (FERC).

Power plants grew larger and larger. A new form of organization arose—the *public utility holding company*. It was an umbrella organization that owned literally hundreds of individual systems. Middle West Utilities Company, Insull's holding company, provided utility services through its operating subsidiaries to more than 5,300 communities in 36 states. In 1932 Insull was president of 11 power companies, chairman of 65 and director of 85. The number of operating utilities dramatically declined. Between 1922 and 1928 the number of individual electric utilities decreased by 33 percent, whereas the number of communities served by the remainder increased by 5,000 or about 37 percent. Between 1917 and 1927, 900 municipal utilities were abandoned. In 1927 only 125 utilities generated electricity for more than 80 percent of the electric customers in the nation. If to this is added the amount of electricity purchased by these utilities for distribution, they supplied almost 97 percent of the nation's electricity. Senator George Norris, father of the Tennessee Valley Authority, proclaimed on the floor of Congress in 1925 that, "I have been dumbfounded and amazed, and the country will be dumbfounded and amazed when it learns that practically everything in the electric world . . . is controlled either directly or indirectly by this gigantic trust."[7]

As one student of public utilities writes:

> "It was a race between the technical achievement of the economies of mass production and the invention of legal devices for mobilizing entrepreneurship to make use of them. . . . By using the devices of the lease, the trust, the corporate merger, and the holding corporation, great pyramids of ownership and control of public utility markets were set up . . . the jurisdiction of state commissions could not reach all the facets of this developing problem. Aggravated by the depression and by the fact that less than half the state commissions had adequate powers over security issues and over mergers and consolidations, the unsound financial structure of many holding companies collapsed in the financial storms which swept the country beginning in October 1929. The administration of President Hoover in Washington temporized with the problem, and hence the control of these "pyramids of power" became an issue in the campaign of 1932."[8]

The New Deal Electrifies the Nation

When Franklin Delano Roosevelt (FDR) was governor of New York, he discovered that the electric bill at his country cottage in Georgia was four times higher than at his home in New York. "It started my long study of proper public utility charges for electric currents and the whole subject of getting electricity into farm homes," Roosevelt later said.[9] To FDR, electricity and development went hand-in-hand. His administration added three federal agencies to the electric system: the Tennessee Valley Authority (TVA), the Bonneville Power Administration (BPA) and the Rural Electrification Administration (REA).

The TVA, created in 1935, is a federally owned corporation for regional development. By 1970 it had become the single largest electric utility in the nation, with twice the installed capacity of any other utility and approximately 5 percent of the nation's total generating capacity.

The Bonneville Power Administration, also created in 1935, is primarily a marketing agency that transmits electricity from federal hydroelectric facilities to investor-owned and public utilities. By 1970 BPA could boast that it operated the nation's largest network of long-distance, high-voltage transmission lines.

In the mid-thirties, the REA offered rural electric cooperatives long-term, low-interest loans for electric generation capacity and transmission and distribution lines. Before REA, many power companies charged rural customers up to 15 times the cost of production. For $5 rural residents could become members of a cooperative or public utility district and own their own power plant or bargain with the previously recalcitrant investor-owned utility for more modest electric rates. The proportion of farms in the United States with electricity increased from 10 percent in 1930 to 43 percent in 1944 to 98 percent in 1975. Rural electric cooperatives, commonly called RECs, now serve 25 million people through 1,000 cooperatives in 46 states. Only a handful of RECs generate their own power (27 in 1974), but they own 42 percent of the electric distribution lines in the nation.

Many of the new utilities were based on hydroelectric power. Back in Teddy Roosevelt's era, the federal government decided that federally owned water resources should be used first to benefit publicly owned agencies. The Reclamation Act of 1906 empowered the Bureau of Reclamation to produce electricity in conjunction with federal irrigation projects and to dispose of any surplus for municipal power. The bureau's role was expanded in the Federal Water Power Act of 1920 and the Flood Control Act of 1944. Both gave a preference for public bodies and cooperatives. This preference clause became important in the 1980s as cities vied with investor-owned utilities to claim the rights to harness hydro on federally owned land. BPA was one marketing agency for federal hydroelectric power. To a lesser degree, the Southeastern, Southwestern and Alaska Power Administrations later played this role.

FDR brought electricity to areas of the country which previously had none. To do so, his administration created new organizational forms. Also on his agenda was the need to regulate the private utility corporation more closely, to avoid the abuses of the utility holding companies. The Securities Act of 1933 and the Securities and Exchange Act of 1934 established the Securities and Exchange Commission (SEC). The SEC had three main operating divisions, one of which was a Public Utilities Division. It had jurisdiction over the issuance of all securities to be sold in interstate commerce, including those of a utility and of a nonutility character.

The Public Utility Holding Company Act of 1935 provides for the physical disintegration of holding company systems and restructures the public utility industry.

These laws did not pass easily. After two stormy years the Public Utility Holding Company Act passed by one vote. Passage was assured only by eliminating the most controversial provisions. For example, bowing to pressure from utilities, Congress agreed not to convert electric utilities into "common carriers." Federal regulatory agencies were therefore denied the authority to order utilities to transmit electricity from an independent power producer to another buyer.

This concession was to return to haunt small power producers 50 years later. A common carrier is a monopolistically owned distribution system that must carry the goods of independent companies. For example, a highway is a common carrier. So are railroads and oil pipelines (although natural gas pipelines are not). If the electric grid system were to be a common carrier, the utilities would have to transmit anyone's electricity at a nondiscriminatory rate. Congress initially ordered them to do so, but then pointedly withdrew that provision. The courts repeatedly referred to this refusal by Congress in their decisions to deny state regulatory commissions the authority to order utilities to transmit electricity from one independent producer to some remote buyer. In the 1960s and 1970s the primary parties hurt by these rulings were the municipal utilities. Having given up their generation capacity because it was cheaper to buy into larger, privately owned systems, they found themselves unable to switch to cheaper producers because the utilities refused to transmit that electricity (this is called wheeling) over their grid system. In the 1980s this lack of common carrier status would inhibit all small power producers from getting the best price for their electricity by eliminating the possibility that they could sell to remote customers.

Power Pools and the National Grid System

The New Deal extended electric power to the entire country, rural as well as urban. It also substantially changed the organizational form of the corporations that generated and delivered electricity. But the technological underpinnings of the electric system remained inexorable. The dynamics of bigness continued to unfold.

Several utilities interconnected in what became known as regional power pools. The first one was established in 1927 when the Public Service Electric and Gas Company of New Jersey (PSE&G) and the Philadelphia Electric Company joined forces. By 1970 there were 17 power pools, representing 50 percent of the nation's generating capacity. By the late 1960s one utility expert could write, "The United States is already close to being a two-network country, and the process of interconnections across the Rockies to link the two networks has already begun."[10]

The interconnectedness of grid systems represented an increasing interdependence. Utility historians like to recall the story of an Ohio utility that suffered a service interruption during the 1960s. It was connected to a regional power pool. The electrical impulses set up by the failure were felt at progressively greater distances as each installation down the line had no available power. The first plant to respond to the need was an idling hydroelectric plant in Arkansas. When the demand reached it, the plant automatically started up. Its gates opened and a large volume of water was released below the plant. At that moment, a man was fishing in a boat too close to the plant, and the sudden rush of water capsized his boat. Utility operations had become so interrelated that a power outage in Ohio could cause a drowning in Arkansas.

Transmission lines were built to carry higher and higher voltages. From 1900 to 1950 the maximum AC voltage transmitted increased from less than 50,000 v to 230,000 v. In the late 1950s, 345-kv lines were in operation, and by 1980 there were 765-kv lines. These transmission lines became super highways for electric power. Each time the voltage was raised, the amount of traffic the line could carry went up. A transmission line rated at 500 kv typically handles about 2,000,000 kw, or 2,000 megawatts (Mw), the output of two giant generating plants. A 765-kv line handles about 3,000 Mw.

Higher transmission voltages went hand-in-hand with larger power plants. The largest steam power plant built in 1952 had a capacity of 125 Mw; in 1967 the largest was 1,000 Mw. On the average, unit size increased by more than 700 percent from 1947 to 1967—from 38 to 267 Mw.

In 1977 there were more than 4,000 power plants in operation, yet fewer than 300, or 7 percent, generated more than half the nation's power. The Federal Power Commission confidently predicted the trend toward bigness would continue. It foresaw 2,000-Mw fossil-fueled plants by the 1980s and 3,000-Mw plants by 1990. A single power plant would be able to serve a city the size of Houston!

Large power plants, with their cheaper power, convinced cities and rural cooperatives to abandon their own capacity and buy into larger systems. In 1935 almost half the municipally owned electric utilities generated all of their own power. By 1975 only one in ten did so.

In 1978 the United States electric utility industry nominally consisted of 3,500 systems, but 2,400 of them were involved solely in the transmission and distribution of power. The combined output of municipalities, public-utility districts and state power authorities accounted for less than 10 percent of the country's total generating capacity. Two hundred and fifty investor-owned utilities owned more than 80 percent of the nation's generating capacity. The top ten of this group owned almost half of this. The day of the small owner-operated power plant appeared to be over forever. Less than 5 percent of the nation's electricity was generated on-site in 1975, and all of this came from large industrial generators.

For all practical purposes the nation was divided into three separate power supply regions: Texas, the eastern states and the western states. The average kilowatt of electricity traveled 220 miles, the distance from New York to Washington, D.C. Electricity generated in British Columbia traveled as far as to southern California and Arizona, while some eastern Canadian electricity probably went nearly to Florida.

The nation continued to find new ways of using electricity. Streetcar companies gave way to industry as the major user by 1920. By the 1960s, residential and commercial buildings were the major consumers of electricity for space heating and cooling. A larger and larger portion of our primary fuels (coal, oil and gas) was being burned to generate power. In 1930, 10 percent of our fuels were used to generate electricity; in 1960, 20 percent and in 1980 almost 30 percent were used for this purpose.

Financial advisors recommended utility stocks for those who wanted a good return with no risk. It was especially attractive for the elderly and pension funds. This was the golden age.

The Golden Age Ends

High voltage transmission lines and interconnected power pools increased the electric system's complexity to an unprecedented level. Scientists and engineers began to encounter strange resonances throughout the system, a behavior and response pattern that could not be explained by existing theories. An entirely new science was needed to understand the new electric synergy.

After World War II, the utilities assessed their ratepayers to finance a new research and development (R&D) organization. The new R&D firm, the Electric Power Research Institute, now manages more than a thousand projects in all aspects of electric energy generation, delivery and use, with the actual R&D going on in industries, utilities and universities.

Power pools were initially justified as a way to reduce reserve margins, but individual utilities refused to view the pool as a reliable backup source. Each utility built its own back-up source. During the period that power pools proliferated, the reserve margins actually increased by almost 300 percent. Recommended reserve margins increased by 100 percent.

Now, however, the cost of transmitting power was becoming a significant factor. By 1972 the cost of building and maintaining the grid accounted for 70 percent of the cost of delivered electricity. We were paying twice as much to get the electricity from the plant to us as we were to get it generated.

The economies of large power plants proved illusory. After 1960 large power plants actually became less efficient. The larger the plant, the more it broke down. Coal plants of from 400 Mw to 800 Mw were inoperable about 8 percent more than plants half as big. For all coal- and oil-fired power plants in the United States during 1967 to 1976, the *forced outage rate* (the fraction of time a plant is involuntarily out of service) ranged from a tiny 2.5 percent for plants under 100 Mw to 16 percent for plants of 800 Mw, rising proportionately in between.

The complexity of the grid system continued to plague its originators. In 1965 a cascading power failure originating in a relay that malfunctioned in Canada interrupted the electrical supply of most of the northeastern United States. Thirty million people lost electric power for up to 13½ hours. Fully 23 percent of the 1965 peak electric demand in the United States was unfilled. A decade later, on 13 July 1977, just three days after the chairman of Consolidated Edison (Con Ed) of New York had said he could "guarantee" that a recurrence was remote, nearly nine million people were affected by a blackout, this time for as long as 25 hours. The assistant director for systems management of the Department of Energy (DOE) noted in 1976, "It is becoming apparent that the increasing complexities of the nation's electric system are rapidly outstripping its capabilities . . . . There does not exist any comprehensive applicable body of theory which can provide guidance to engineers responsible for the design of systems as complex as those which will be required beyond the next generation."[11]

Amory and Hunter Lovins, after an exhaustive analysis of the weakness of our electric transmission system, concluded in 1981, "We may well find, as power systems evolve in the present direction, that they have passed unexpectedly far beyond our ability to foresee and forestall their failures."[12]

The transmission systems increasingly became the soft underbelly of the electric system. Of the 12 worst power interruptions to the bulk power supply in the United States from 1974 to 1979, 6 were caused by failures in transmission, 6 in distribution and none in generation. Seven were initiated by bad weather, 4 by component failure and 1 by operator error. Sometimes the most minor mishap ended in widespread disaster. On 8 January 1981 a trash fire at the Utah State Prison apparently caused arcing in a major switchyard next door. The resulting quadruple transmission failure blacked out all of Utah and parts of Idaho and Wyoming. One and a half million people were affected.

Some observers have worried about the possibility of sabotage. The Government Accounting Office audited the electrical security of a typical part of the United States and determined that the sabotage of only eight electrical substations could black out an entire region. The sabotage of only four substations would leave a major city with no power for days and with rotating blackouts for a year.

The increasing separation of production and consumption in the electric system undermined the ability of communities to control their futures. Communities at different ends of the "electric pipeline" fought one another. For example, in western Utah around Delta, construction of the largest coal-fired facility in history began in the late 1970s. Electricity from the 3,000,000-kw facility would be transmitted 500 miles to southern California. Steam plants need water. Water is scarce and precious in Utah. The facility, 50 percent owned by California municipal utilities, bought up 40,000 acre-feet of water in 1981. To air-condition Los Angeles, the economy of western Utah was going to change from agriculture to mining.

The separation of the generation facilities from the final consumers meant the costs and benefits of electric power were imposed on different communities. While one community fought the disruption that came with new power plants, another community basked in increased electric capacity and increased its demand accordingly.

High-voltage transmission wires require wide rights-of-way. Using their power of eminent domain, utilities expropriated wide swaths of private land to erect the six- and seven-story towers. People fought against this intrusion on their property and against possible harm from the magnetic fields emanating from the 765-kv lines. Bitter confrontations took place from 1979 to 1980 between Minnesota farmers and utility companies trying to build these lines. Eight thousand fragile glass insulators were shot out by rifles. Guarding just the Minnesota section of line required 685 watchtowers spread over 176 miles. "Despite high-speed helicopters, a reward of one hundred thousand dollars, three hundred private guards and extensive FBI activity, not one of the perpetrators has been caught. It is not likely that they will be, given the depth of their local support,"[13] wrote the Lovins.

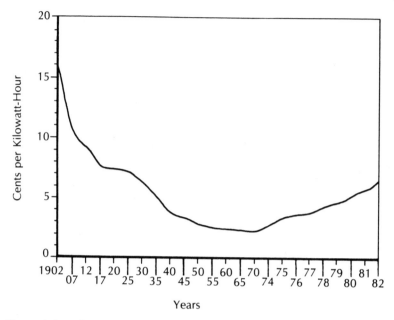

Figure 1–3: This graph shows energy cost changes from 1902 to 1982.

The death knell of the golden age of electric power came with the dramatic rise in fuel prices in the 1970s. By 1970 almost 40 percent of the nation's generating capacity was oil-fired at a time when a barrel of oil cost $1.75. By 1980 that price had risen to $33. The deregulation of natural gas would bring its price up to that of oil by the mid-1980s.

For the first time in a century, electric prices rose dramatically. The average cost per kilowatt hour in 1907 for the residential customer was 10.5¢. In 1970 the average price was 2.1¢. The average factory worker earned 20¢ an hour in 1907 and $3.36 an hour in 1970. Therefore, to keep a 100-watt light bulb burning all day, the laborer in 1907 had to work 30 minutes. The laborer in 1970 had to work 23 seconds to buy the same amount of electricity!

Regulatory commissioners, whose easy task had been to determine how fast prices should drop, now had to impose higher tariffs. Demand no longer rose at the historic 7 percent annual rate. In 1977 it rose by 4 percent; in 1980 by less than 2 percent. Instead of doubling every 10 years, demand was doubling every 35 years. In fact, growth continues to slow down. From January 1981 to January 1982, the total electricity sales in the country increased by only ½ of 1 percent. At that rate electric demand would double by the year 2126!

Demand forecasting has indeed proved to be a risky process. From 1973 to 1982, the Edison Electric Institute, the nationwide association of investor-owned utilities, overestimated projected demand by more than 100 percent every year. But even as forecasting was revealed to be an inexact science, the penalties for guessing wrong have become more severe. Nowadays, the bigger the power plant, the longer it takes to come on-line. By 1982 the size and complexity of nuclear power caused a 12-year delay between the preliminary planning and the actual generation of power. Power plants conceived (by projected growth rates) in 1970 weren't born until 1982. But during those 12 years, projected demand increases had fallen short by up to 90 percent. In most parts of the country, far more capacity was coming on-line than was needed to meet demand, even on the hottest summer day or coldest winter night. Electric rates had to increase to pay for these increasingly idle power plants. Meanwhile, the cost of idle plants rose. Utilities, used to paying 1 percent interest in the early 1960s, were forced to borrow money at the 15 to 18 percent interest rates of the early 1980s. Nuclear power plants built for $400 per kilowatt in 1970 cost $2,500 per kilowatt in 1982.

By the 1980s, the amounts that utilities were spending on a single power plant were astonishing. Building one nuclear plant could double a utility's entire previous investment in facilities. "We've never seen lumps like this in the past," said Alfred E. Kahn, a specialist in regulatory economics and former special advisor to Jimmy Carter. To pay for such plants, utilities have begun to ask for sharp rate hikes, but these hikes can dampen demand even further. One Long Island Lighting Company (LILCO) vice-president reflected in mid-1982 that "you have the problem of prices cutting into sales." LILCO asked the New York Public Service Commission for permission to phase in rate hikes over several years instead of all at once, even though doing so might alienate investors. Phasing in construction costs, in the opinion of Dan Scotto, vice-president for electric utility companies at Standard & Poor's Corporation, would mean that bond ratings would be lower. Lower bond ratings mean higher interest costs and, ironically, still higher electricity prices.

Meanwhile, the demand for electricity has continued to drop. TVA has canceled a half-dozen proposed nuclear plants. Utah Power and Light has recommended deferring half of the proposed capacity for the Intermountain Power Project outside of Delta. The Washington Public Power Supply System has admitted that two and possibly three of its nuclear plants under construction would not be needed for the fore-seeable future.

The actual decline in electric demand in the early 1980s was largely a result of the economic recession that plagued the world in the aftermath of the rapid oil price increase of 1979. If the economy recovers, it is likely that the demand will once again rise. However, the rate of increase is unlikely to ever again reach the lofty previous levels. Efficiency and the substitution of other types of energy for electric energy have become very cost competitive. For example, General Electric's electric motor factory in Erie, Pennsylvania, is gearing up for an economic recovery by increasing its production several-fold. The new motors will use 40 percent less electricity per mechanical energy produced than did their predecessors. As industry upgrades its industrial processes, it will be buying newer, more energy-efficient equipment. Residents will be trading in their old cars and old appliances for more efficient ones.

The Pendulum Swings Back: Decentralized Power

The trend toward centralization reversed itself in the 1970s. Once again small dispersed power plants of less than 10 Mw became economically attractive. One reason was that smaller plants could come on-line rapidly, often in fewer than three years. Thus investments in additional capacity could be more easily matched to changes in demand. Forecasters would no longer have to bet billions on ten-year projections. Also, short-term financing in the uncertain capital markets of the 1980s is much easier to obtain than the 20- and 30-year bonds utilities have had to issue to finance large central power stations.

Large numbers of small power plants give the electric system greater reliability. Since they can be located nearer to the final user, the transmission and distribution costs can be reduced. Also, smaller plants lend themselves to mass production techniques that can lower unit costs.

New and refashioned electric generation technologies have entered the marketplace. For example, *cogeneration,* which has been in use since the last century, is now proven to be economical on a much smaller scale than was previously the norm. The same is true of hydropower systems: smaller is more profitable. Other technologies, such as wind power and photovoltaics, have also made gains with recent advances in electronics and materials sciences.

Cogeneration is a process by which systems produce both electrical (or mechanical) energy and thermal energy from the same primary energy source. Conventional energy systems supply either electricity or thermal energy, while a cogeneration system produces both.

A typical commercial boiler that is used to heat an apartment house or business complex has an efficiency of about 50 percent. More than half the energy in the fuel is wasted. A typical central power plant has an even lower efficiency, in the range of 33 percent. Subtract from that additional losses in transmission, and almost three-quarters of the energy burned in a power plant is lost before the electricity enters the building.

Cogenerators, on the other hand, have efficiencies of 75 to 95 percent. These efficiencies can be achieved only if a nearby use can be found for the waste heat. Thus cogeneration units are usually placed inside or near the buildings to be served. A 1978 study by the State of New Jersey discovered that 50 percent of the boilers in state buildings were over 25 years old and would soon need to be replaced.

The study recommended they be replaced with cogenerators that would increase their efficiency by 50 percent, generating electricity as well as thermal energy. In the early 1980s, several automobile and truck companies were redesigning their basic engines into household-

Photo 1–2: This single-family residence in Carlisle, Massachusetts, generates an annual surplus of electricity from its 7.5 peak kilowatt rooftop photovoltaic array. Solar Design Associates, Lincoln, Massachusetts, were the architects and engineers. *Photograph courtesy of Solarex Corporation.*

sized power plants by linking them up to generators and installing heat recovery equipment.

The benefits of small cogeneration systems are not restricted to plants that are fueled by gas or oil. Canada is developing a nuclear-powered cogeneration plant. Atomic Energy of Canada, Limited, is developing 2-Mw to 20-Mw units that could heat and power a small city. The plants would be unattended most of the time, responding automatically to daily variations in demand. The reactor core would contain enough uranium fuel to last two heating seasons. The Canadians are attempting to uprate (increase peak output) a 20-kw research reactor called SLOWPOKE, developed in 1970. The researchers conceded that the "public may not readily accept small nuclear reactors in place of oil furnaces," but they believe in the inherent safety of these miniature plants. "A decentralized system of small reactors, which effectively eliminates the possibility of a single big accident, may have significant advantage in licensing, insuring and gaining public acceptance. Eventually the public may accept accidents to small reactors to the same extent that they accept fires, explosions and air-crashes. . . ."[14]

Out of the reversals in the trends of energy costs, growth and demand, a new industry is being born to deliver the necessary goods and services for small-scale power production. Small businesses have developed prototypes, worked out the bugs, retooled and evolved reliable machines. Every increase in the price of conventional electricity has made solar power plants economically attractive in a wider range of locations. Hydroelectric plants were common a century ago. But by the 1950s, the price of oil was so low that, in order to be competitive, only huge systems could be built on the largest rivers. But as the price of oil has risen, smaller-scale facilities have again become attractive. Towns that had abandoned their turbines in the 1940s and 1950s have begun to refurbish them. In 1979 the Army Corps of Engineers identified more than 3,000 potentially economical hydro sites on existing dam sites alone. These sites could generate economical electricity for several thousand homes. The term *small-scale low-head hydro* entered the energy vocabulary. Today's prices for conventionally generated electric power have increased sufficiently so that even minor and slow-moving rivers and creeks could be economically harnessed. The term *micro hydro* was coined and quickly adopted. These systems are economical even if they serve only a few homes.

The wind power industry has evolved with equal dynamism. To be competitive with diesel generators in the 1950s, wind turbines would have required wind speeds higher than those in any part of the nation. By 1982 average wind speeds of less than 15 miles per hour (MPH),

Photo 1–3:   A 3-kw wind machine is pictured here. With adequate winds it could supply at least one half of an average residential electric load, not including electric space or domestic water heating. *Photograph courtesy of Joe Carter.*

which are available in significant portions of the country, could generate electricity competitively from some of the new generation models developed in the seventies.

The most dispersed of all energy sources, direct sunlight, has proved to be a strong competitor in the very near future. *Photovoltaics,* involving the use of *solar cells,* had only been in existence since the early 1950s. They were used only to power satellites until 1973, but in 1974 the first manufacturer of cells for terrestrial applications set up business. At that time the price for photovoltaic power was more than 300 times that of conventional power plants, but by the late 1970s the price had dropped to where it was 50 times more costly. By 1982 the cost of photovoltaic electricity was only 10 times that of conventional electric power. Although it is still too expensive for widespread applications, almost 1,000 households are using solar cells. They are mostly used on homes located off the grid system. Compared to the cost of laying miles of electric cable to connect to the grid, the solar cells proved to be more economical. The photovoltaic industry has predicted that, by the mid-1980s, its products will be competitive for grid-connected applications in most parts of the nation.

The increasingly marginal economics of modern electric power plants and transmission systems has encouraged businesses to design technologies that could operate efficiently in dispersed arrays. But the new breed of electric producer threatened the existing utility structure. As the Congressional Office of Technology Assessment concluded in 1978, "If energy can be produced from on-site solar energy systems at competitive prices, the increasing centralization that has characterized the equipment and institutions associated with energy industries for the past thirty years could be drastically altered; basic patterns of energy consumption and production could be changed; energy-producing equipment could be owned by many types of organizations and even individual homeowners."[15]

Utilities have worried about the fragmentation of the electric system. To them it represented a regressive tendency. Thomas Hurcomb of Central Vermont Power expressed such concerns before Congress in 1978. He warned, "If we continue to break down . . . we come up with what we had back 50 or 60 years ago of a hundred or more utilities. . . . I believe that will make the planning process more difficult. I believe it will make energy more expensive. I do not think that the course that we should be following is continually to break down into smaller energy groups."[16] Utilities certainly had the means to delay significantly the proliferation of independent power systems. They controlled the grid. They had no responsibility to interconnect with the small power producer. They could, and did, charge very high prices to those they allowed to interconnect, thereby forestalling potential future interconnections.

Under existing law, there was little the state regulatory commissions could do to aid the independent producer. A survey of the 50 regulatory commissions in 1978 found the vast majority believed they lacked the legal power to order utilities to interconnect and they could not require utilities to buy power from independent producers.

In 1978 Congress made a landmark decision that resolved the dilemma and opened the floodgates of independent power production. The Public Utility Regulatory Policies Act (PURPA) abolished the century-old monopoly utilities had over power generation. To reduce the nation's dependence on imported oil and increase the efficiency with which electricity is generated, PURPA required utilities to interconnect with qualifying facilities and to purchase power from them at premium rates. The act applied to all utilities, whether investor owned, cooperative or municipally owned. And it exempted these new producers from state or federal utility regulations.

The passage of PURPA and the coincident enactment of tax benefits for cogeneration and renewable energy electric plants created a new industry almost overnight. Investors quickly rushed in to buy up the windiest terrain and the best hydropower sites. Journalist John McPhee described the excitement in the small hydro market in the *New Yorker* in 1981. "It is possible that in 1897 less action was stirred by the discoveries in the Yukon. There was a great difference, of course. The convergence of the Klondike was focused. This one—

Photo 1–4: This owner-built paddle wheel is the heart of a small-scale hydroelectric system. The paddle wheel is designed to work in this low head application where most of the power is gained from the rate of flow rather than from water falling to a lower level. *Photograph courtesy of Tanya Berry.*

this modern bonanza—was diffused, spread among countless localities in every part of the nation. As a result it was a paradox—a generally invisible feverish rush for riches."[17] Wind prospectors scoured America's windy coastlines and plains.

Applications for licenses to refurbish existing hydroelectric sites poured into the FERC. Enterprising companies have recently set up hundreds of small wind turbines in densely packed arrays. The nation's first wind farm started operating on a New Hampshire hilltop in late December 1980 with ten machines of about 50 kw each. By mid-1982 seven more wind farms were operating, primarily in California. As the Idaho Public Utilities Commission noted in August 1980, "No longer is [electric generation] to be the exclusive domain of public utilities. Their natural monopoly has always been and will continue to be the distribution of electricity. Henceforth, however, electric generation is to be a competitive enterprise with regulation intervening only to the extent necessary to stimulate a free market."[18]

The transition is not going to be an easy one. Upon the enactment of PURPA, many utilities immediately filed suit to overturn the legislation. In March of 1981, the same month the PURPA regulations were to go into effect, Judge Harold Cox of the Southern District Court of Mississippi upheld the contention of the Mississippi Power and Light Company, the state of Mississippi and the Mississippi Public Service Commission in declaring PURPA unconstitutional. "The sovereign state of Mississippi is not a robot or lackey which may be shuttled back and forth to suit the whim and caprice of the federal government," he ruled.[19] In the spring of 1982, by one vote the United States Supreme Court overruled Cox. PURPA stands.

The industry of dispersed power production is still embryonic. Yet at its present rate of growth, it threatens soon to surpass investments by utilities in conventional power plants. In 1982 investor-owned utilities spent about $25 billion for generation capacity. The Edison Electric Institute predicts this investment will shrink to less than $15 billion in 1986 (in 1982 dollars). On the other hand, the FERC reports that filings from potential qualifying facilities (QFs) under PURPA have risen from 30 in 1980 to more than 500 in 1982. The 500 plants proposed in 1982 have a combined capacity of more than 11,000 Mw. Assuming an average investment of $1,000 per kilowatt of installed capacity, this will represent an $11 billion investment. All that investment will not be spent in one year, but disbursed over three years. Thus $4 billion will be invested in nonconventional power plants in 1982, 15 percent of the utility total. By 1986 investments in cogeneration and small power production facilities could exceed those by conventional utilities in traditional power plants.

Photo 1–5: Cogeneration is exactly what the name implies: a system that simultaneously produces both electricity and heat from the same primary source of energy. Early in this century, cogeneration systems provided over half of the energy used in the United States. But the technology fell into disuse because of cheap oil prices and the rise of modern central electric utilities. Now, because of ever-increasing oil prices and the high cost of building new central power plants, cogeneration, which can yield efficiencies of 75 percent and more, is once again economically attractive. *Photograph courtesy of Agway Research Center.*

These phenomenal increases are a direct result of PURPA and ensuing state legislation. In 1979 utilities in New Hampshire were paying an average 2¢ per kilowatt-hour for electricity generated by independents. That year the state legislature set the price at a minimum of 4¢. In 1980 the public service commission raised that minimum to almost 8¢ per kilowatt-hour. In Montana, the first contract signed under PURPA regulations included a 3¢ per kilowatt-hour price. In 1982 Montana raised the minimum to 6¢. The 1982 New York State legislature mandated a minimum 6¢ per kilowatt-hour rate pending a public service commission investigation of whether higher rates were warranted.

Meanwhile, new trade associations have been formed. The American Wind Energy Association, the National Alliance of Hydroelectric Enterprises and the Cogeneration Coalition were formed from 1979 through 1981. The California Independent Energy Producers Association brought all technologies under one umbrella in mid-1982.

PURPA was designed to encourage competition in power generation. However, it retained the utilities' monopoly over transmission and
distribution. Even as the 50 state regulatory commissions and thousands of business corporations and cities were working out ways to
disperse generating capacity, other groups were exploring the next
step. At the Massachusetts Institute of Technology (MIT), the Homeostatic Energy Group explored the feasibility of transforming the grid
system into a giant marketplace. The small power producers would sell
their electricity to the grid as if it were a brokerage agency. As demand
and supply fluctuated, the price of electricity would also fluctuate.
Electricity sold in the morning hours in areas with low demand would
receive a low price. Electricity sold during summer afternoons in places
with a high air-conditioning load would receive a high price. Micro-
processors would record all transactions and establish prices on five-
minute intervals.

In early 1983 the Pennsylvania Electric Utility Efficiency Task
Force recommended experiments to open up the grid system to what
it called "self-help electricity." Self-help electricity programs would
allow consumers to contract for power directly with independent electricity sources. Self-help customers would use their local utilities only
to carry the contracted electricity over the utilities' transmission lines.

It is important to keep in mind that PURPA was not enacted to
promote small power production. It was enacted to reduce dependence on foreign oil and to increase efficiency in generating electricity.
The benefits of PURPA are available to any producer that uses renewable resources in a power plant as large as 80,000 kw. Cogenerators
have no size limits at all. The size limit of 80 Mw may be small by
investor-owned utility standards, but it is not small by the standards of
most municipal utilities or rural cooperatives. For these utilities, the
irony is that PURPA could actually encourage more centralized electric
power production. One could imagine a cogenerator of 300 Mw
swamping a small municipal utility with two 80-Mw power plants.
Already a small 20-Mw nuclear reactor, at Argonne National Laboratory, has become a qualifying facility under PURPA. Since PURPA
basically eliminates regulatory oversight for these facilities, there is the
potential that a system that is dominated by a few regulated companies
will be traded in for one dominated by a few unregulated companies.
This is one reason this book discusses only facilities with less than
200-kw capacities. Another reason is that PURPA makes a distinction
between those facilities with more or less than 100-kw capacity. A
standard tariff must be offered those under this size. Those above 100
kw are usually required to negotiate individual contracts. By choosing
the 200-kw limit, both cases are covered. This limit also allows the

book to go beyond the individual household application to include small commercial and apartment house facilities. The rough rule of thumb is that a kilowatt of capacity serves one person. Therefore, a 200-kw power plant can serve a small apartment building or a commercial complex or a nursing home or motel.

Finally, there are valid arguments that facilities of less than 200 kw present different burdens and benefits on the electric system from those of the 50-Mw to 300-Mw range. Certainly a utility with a 100-Mw average load can argue that a 200-Mw power plant can, in fact, unbalance and make less reliable its electric system, unlike a series of dispersed 10-kw to 100-kw power plants. The standards for interconnection should also differ considerably for small and larger plants.

The New Power Producers

As one might suspect, the first owners of small power systems come from many backgrounds. Yet they possess two common characteristics: a strong entrepreneurial drive and an ability to understand electrical circuits.

Ted Keck, the 37-year-old owner of a 70-kw hydroelectric facility at a once-abandoned mill site in Pillow, Pennsylvania, learned electronics as the owner of a theatrical lighting company. He used to live 14 miles from the Three Mile Island nuclear facility. The near-meltdown there catalyzed his investigation of alternative energy. Beginning with a solar greenhouse, he eventually explored the feasibility of producing electric power beyond his own needs. He sold his lighting business and moved to the mill.

Joseph Ellen is the owner of a 180-kw hydro facility on an existing dam in the Piedmont section of North Carolina. He is an industrial electrical contractor and, having worked with utility engineers "my whole career," he encountered few problems working out interconnection standards for his facility. Bruce Sloat owns three hydro facilities in New Hampshire. He is both a farmer and a master electrician. Pentii Aalto, a mechanical engineer, owns a 5-kw diesel (oil-fired) cogeneration system in his basement in Braintree, Massachusetts. Bill Clayton owns an 80-kw wood gasifier cogeneration system in Huntsville, Alabama. He is an electronics engineer who designs microcircuits.

Those with technical expertise and a curiosity about independent power production may indeed be the first ones in the water. But hard on their heels are a second generation of pioneers. Ernest L. Copley III, owner of a 15-kw photovoltaic system in Denton, Maryland, is a broker for E. F. Hutton. His facility has only one function: to feed electricity into the grid. Copley views it solely as an investment vehicle and chose to site it in Denton because the local utility pays the highest PURPA rates in Maryland. He is already negotiating for a second qualifying facility in Florida. Victor Lund has installed a 75-kw, gas-fired cogeneration system in one of the eight hotel-like retirement homes he owns in Escondido, California. A $6,000-a-month electric bill made him look for a better investment. H. L. Ayers owns three 20-kw wind turbines near Crowell, Texas, and is a full-time farmer growing wheat, cotton and alfalfa.

The pioneers' motivations vary. Some, like Copley, are attracted by the investment potential. Others, like Sloat, view hydro power as another "cash crop" to be harvested along with apples and vegetables. Still others, like Lund, worry about the impact of rising energy bills on his senior citizens. Most like the feeling of achieving a certain self-reliance that comes from having an independent source of electric power.

Photo 1–6: Wind farms like this one in the Altamont Pass, about 45 miles east of San Francisco, are a portent of the future when there will be thousands of energy farms throughout the country producing electricity from wind, solar, cogeneration and hydro power. The wind machines pictured here are among the first of some 500 that one company is installing under contract with Pacific Gas and Electric. The turbines are mounted on 40- and 60-foot towers. The rotors are about 32 feet in diameter, and each machine has a peak output of 80 kilowatts. *Photograph courtesy of Pacific Gas and Electric Company.*

Be Your Own Power Company

This book is written as an aid to understanding the new age of electric power. It is not so much a how-to manual as a primer on utility economics that emphasizes on-site power generation. At present the negotiation process between the independent power producer and the utility is lopsidedly in favor of the utility. This book is intended as a step toward redressing that imbalance by providing information and a conceptual framework for those who desire to become more independent and/or to use their ability to generate electricity to gain a source of revenue.

Four technologies are discussed: cogeneration, wind power, hydropower and photovoltaics. Each technology is discussed and evaluated in terms of grid-connected or stand-alone power generation.

The present utility system is not easy to understand. As with all industries, the utilities have their own jargon. Certain electrical concepts, such as harmonics, are still not clearly understood even by learned electrical engineers. But, ready or not, the nation is plunging into one of the most dramatic structural changes in its history. This book is intended to aid those desiring to understand these changes and to be part of the changes themselves.

CHAPTER **2**
How the Electric System Works

The millions of miles of wire that comprise the electric grid system can be likened to the thousands of miles of capillaries, veins and arteries in the human body. When you exercise a muscle, blood flows to that area of the body. Similarly, when an electrical device is turned on, electricity flows to that "load." Exercise a muscle more strenuously and more blood is drawn to the area, moving in great torrents through several major arteries to dozens of smaller vessels and thousands of even narrower capillaries. The "load" can be traced back to, or as students of electricity say, "seen at," the body's power plant—the heart.

A spidery web of electrical wires crisscrosses the nation. More than 365,000 circuit miles of overhead high-voltage transmission lines carry huge amounts of power to load centers in the United States. More than 4 million miles of distribution lines carry lower voltage electricity to customers.

Massive trunk lines carry vast quantities of high-voltage electricity over hundreds and even thousands of miles. At substations a device called a *transformer* raises (steps up) or lowers (steps down) the voltage. Typically the voltage is raised 1,000 volts (v) for every additional mile electricity is transmitted. Large power plants feed power into bulk transmission lines 10 to 20 miles away at 13.8 kilovolts (kv). A transformer steps up this electricity to voltages of 69 kv to 745 kv, depending on the length of its journey through the wires. The 270-mile transmission line from the Hoover Dam to Los Angeles carries 275 kv. When the high-voltage transmission line reaches a load center, subtransmission transformers step down the voltage to between 69 kv and 138 kv, with the latter voltage being sufficient to send electricity about 110 miles. Some very large industrial customers can make direct use of these high voltages. They may have a single transformer for their own electricity, called a *dedicated transformer*.

Distribution substations are often located in or near towns. From the distribution substation, the electricity flows into feeder lines. Each substation may supply up to 25 feeders. The *feeder primary* is the main

35

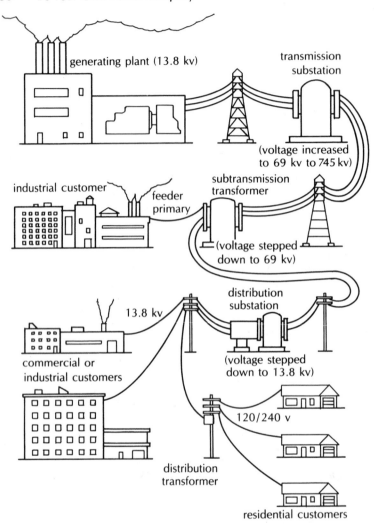

Figure 2–1: The voltage leaving the generating plant is raised (stepped up) by a transformer to very high levels so that electricity can be transported long distances with few line losses. The voltage is lowered (stepped down) by transformers several times to meet the needs of different customers.

line carrying power from the distribution substation. It operates at voltage levels of 2.4 kv to 32 kv. The particular voltage level of a feeder primary depends on the power demand in the feeder's service area. For example, a 15-kv feeder primary will have a peak capacity of 5,000 to 10,000 kilowatts (kw). This size feeder could serve approximately 2,000 homes.

Once again the voltage levels of the feeder primary can serve some industrial customers directly. Electricity sold at higher voltage is cheaper for the utility to deliver, so the customer will be able to buy it at a lower price. Branch lines called *laterals* radiate from the feeder primary. Some commercial customers are connected directly to these branch lines. Voltage must be reduced further through a distribution transformer connected to residences. Each distribution transformer serves one to ten homes. The lines on the customer's side of the distribution transformer are known as the *distribution secondary system*. All houses within the same distribution secondary system are connected in parallel with each other. Thus a short circuit in one person's home will not cut off electricity in another's. But these houses that share a transformer do represent a sort of mini-electrical community. The efforts of one house to move toward electrical self-reliance can, in fact, have an effect on the other houses in the distribution secondary system. As will be discovered, one issue raised by utilities when interconnecting small power producers to the grid system is whether they should continue to share a distribution transformer with other customers or be required to install a transformer solely for their own use.

The distribution secondary in residential areas carries the standard 120/240 v, single-phase power. Single-phase power consists of one signal. In contrast, transmission lines and distribution primary and lateral lines carry three-phase power, which consists of three signals (see figure 2-2). Each signal is out of phase with the other two by 120 degrees. Three-phase power is used by industrial equipment and larger commercial appliances. Motors and other types of machinery cost less and operate more efficiently when supplied by three-phase power. Moreover, it is more economical to transmit three-phase power than single-phase power.

Electricity "leaks" on the way from the power plant to the ultimate user. The leaks are known as *line energy losses,* or just *line losses.* Two factors account for these losses: distance traveled and resistance to electricity flow. Because transmission lines are high voltage, relatively little energy is lost along them.

This is because for a given amount of power (watts) transmitted, the greater the force driving the electricity (volts), the lower the current (amps). The power lost is given by the formula $P = I^2 R$ (where P is power, I is amps and R is resistance). Thus by dropping the current in half, the line losses are reduced by 75 percent.

Proportionately greater line losses occur during lower-voltage distribution because of the higher current. In all, about 9 percent of the electricity is lost getting from the central power plant to the customer. An additional 2 percent is lost in the transformers.

This vast web of electrical service delivery facilities is expensive. The transformers that step up the generator's output to the high voltages necessary for long-distance transmission cost over $2.5 million each. High-voltage power lines cost about $220,000 per mile. Bulk substations that make the first voltage step-downs run as much as $10 million dollars. The smaller lines bringing electricity to individual homes and businesses are less expensive than the big ones, but still cost about $30,000 a mile. The pole-top, or underground, transformers in your neighborhood cost about $1,000, while the cost of bringing electric power that final few feet into your home costs $150. Typically investments in transmission and distribution (T&D) facilities represent 50 percent of a utility's total system investment.

A utility transmission and distribution system includes a vast array of components besides lines and transformers. Substations in the system contain voltage regulators, circuit breakers, protective relays and other switching equipment. There also are relays, fuses and sectionalizing switches throughout the distribution system. This equipment helps maintain acceptable voltage levels and provides protection against excessive currents, unbalanced voltages and currents between phases, and other abnormal conditions known as faults. When a fault is detected, circuit breakers and fuses open in order to isolate the malfunctioning circuit. In some cases, the circuit breaker or fuse is equipped with a recloser to clear temporary (called transient) faults such as those that occur when two lines momentarily touch.

The electrical grid system is awesome in its complexity. Yet it is not perfect. Despite the best efforts of modern electrical engineers and system designers, things can and do go wrong. One such event occurred the evening of 13 July 1977 in New York City. At the time there was plenty of generating capacity available in the pool of adjacent utilities with which Consolidated Edison of New York (Con Ed) was interconnected. But events developed in such a manner that there was no way to deliver that power to the city. The grid system didn't work. Philip Boffey provides an excellent description of the failure sequence and a clear warning against engineering hubris in the following excerpt from his story in *Science* magazine:

> *The trouble began at 8:37 P.M. on 13 July when lightning struck a tower [which was imperfectly grounded] in northern Westchester County and short-circuited two 345-kilovolt lines. . . . Once the short circuit occurred, protective relays, the next line of defense, triggered circuit breakers to open at both ends of the affected lines, thus isolating the problem from the rest of the system. That is exactly*

what the circuit breakers are supposed to do. However, they are also supposed to reclose automatically once the fault dissipates, and this they failed to do. One transmission line failed because of a loose locking nut [which released air pressure from a circuit breaker] in a circuit; the other because a reclosing circuit had been disconnected and not yet replaced. . . .

Two other facilities also tripped out of service. . . . A nuclear reactor [Indian Point Three] shut down automatically when the circuit breakers that opened to contain the lightning fault also [by a design fault] deprived the reactor of any outlet for its power—a design feature that has since been criticized by most investigators. And another 345-kilovolt line—a major tie across the Hudson—tripped out because a protective timing device was designed improperly by Con Ed. . . .

Thus, in one stroke of misfortune, Con Ed lost three major transmission lines and its most heavily loaded generator.

Even so, Con Ed regained its equilibrium by importing more power on the remaining tie lines and by increasing its own generation somewhat. . . . Then lightning struck again . . . and short-circuited two more 345-kilovolt lines. Again there was a malfunction. One line reclosed automatically as it was supposed to; the other remained open because a relay had been set primarily to protect a nuclear reactor (which, ironically, was out of service) rather than to facilitate reclosing of the line. . . . The loss of the line triggered still another malfunction; it caused a temporary power surge that tripped out another 345-kilovolt line. This should not have happened but did, because of a bent contact on a relay. . . .

Con Ed's control room succumbed to confusion and panic. . . . [The] system operator [assumed] . . . a particular transmission line was still in service [and] . . . failed to read a teletype [saying it was down]. . . . Moreover, because of Con Ed's antiquated control room layout, he was unable to see a more dramatic indicator in another room—a flashing screen with a high-pitched alarm. The personnel there knew the line was out but failed to tell him.

As it was, he ignored repeated suggestions from the power pool that he shed load. Then, as the situation deteriorated, he essentially abdicated and dumped the decision-making responsibility on his boss, the chief system operator, who sat at home in the dark reading diagrams by a kerosene lantern and issuing orders over the phone. . . . The chief ordered voltage reductions—but they were too little and too late. Eventually he also ordered that a block of customers be disconnected. Whereupon the confused system opera-

*tor pushed the disconnect buttons and nothing happened. Under
stress, he apparently turned a master switch the wrong way.*

*The performance of Con Ed's reserve generators was equally
erratic. Con Ed's system operator delayed 8 minutes after the first
lightning strike before requesting a fast load pickup from generators
that were supposedly able to respond in 10 minutes. He got only half
the power he expected — and only 30 percent of what Con Ed had
incorrectly told the power pool it could provide. Some equipment
malfunctioned; other units were undergoing routine inspection but
had not been removed from the fast-start availability list; some were
not even manned. [All the night shift operators had been sent home,
and the remote-start capability had been removed some years
earlier. At most 55 percent of Con Ed's total in-city generating
capacity was actually operable.] Similarly, when Con Ed sounded
the maximum generation alarm some 10 minutes after the second
lightning strike, it again failed to get the anticipated response from
its 30-minute reserve generators.*

*As the system cascaded toward collapse, heavy overloads
caused the failure or deliberate disconnection of all remaining ties
to neighboring utilities. Con Ed was now an island, isolated from
outside help. Its last hope was an automatic load shedding system
that had been installed after the 1965 blackout. The system worked
beautifully to disconnect customers. . . . But it also unexpectedly
caused a rapid rise in system voltage that caused a major generator
to shut down. That sealed the system's doom. . . . The remaining
generators could not restore equilibrium. Eventually, protective
relays shut them down to prevent damage. By 9:36 P.M. the city was
blacked out.*[1]

The point of this story is that despite all the relays, regulators, fuses
and other fail-safe equipment the utilities use, and despite the use of
very sophisticated computers to monitor their entire system, things can
and do go wrong. Things happen quickly, too. The Con Ed system went
down in just nine minutes. And, given the interconnectedness of the
modern grid systems, when things go wrong the results can affect
millions of people. This story also exemplifies the fact that most blackouts occur because of failures in the transmission or distribution systems, not because of power plant failures.

An interesting footnote to the 1977 blackout is that now whenever
the probability of a thunderstorm rises above a certain level, New York
City is cut off from the upstate transmission system. During storms Con
Ed generates electricity from very expensive gas turbines. Thus, New

York continues to pay millions of dollars each year as a penalty for the 1977 blackout.

Attention is now given to the generation portion of the electrical system, the modern power plant.

The Modern Steam Plant

The vast majority of power plants now operating are steam plants. The equipment that makes electricity is complex, but the theory behind electric generation is quite simple. Imagine a teakettle, a fan, a magnet and some coiled wire connected together. Fill the teakettle with water, and apply heat. When the water boils, it generates steam, which has tremendous power. As it shoots out of the teakettle, it hits the fan blades, turning the fan shaft. The revolving shaft turns the magnet inside the coils of wire. The motion of the magnetic field near the wire moves electrons inside the wire, and this electron movement is electricity.

A giant steam-electric generating plant operates on this same principle. But instead of teakettles, fans and magnets, the plant uses boilers, turbines and generators.

In a steam generating plant, liquid water is boiled by applying heat from some source, such as coal or oil combustion or nuclear fission. In a coal-fired plant, combustion temperatures may reach 2,600°F. The water changes from a liquid to a vapor (steam) and, in the process, increases its volume many times. The steam expands out of the boiler through a turbine and turns the fanlike blades. The steam is superheated, reaching more than 1,000°F. It is then collected and at more than 2,500 pounds of pressure per square inch, shoots into the high-pressure section of the turbine. The turbine shaft turns a generator to produce electricity.

When low-pressure steam leaves the turbine, it condenses back to a liquid by cooling. This occurs in a condenser where the piped steam is brought into contact with cooling water, which draws away heat, causing the steam to condense into water. Since large volumes of cooling water are required, plants are located near large bodies of water. Power plants are capable of raising the temperature of entire lakes and rivers several degrees by using their water for cooling. (Cooling can also be accomplished with huge cooling towers that use massive amounts of air to take away the heat.) The same water is then pumped back into the boiler and the process is repeated.

Steam power plants only achieve a 35 percent overall efficiency. Every plant keeps records of how much fuel is burned and how much electrical energy is generated over a period of time. The ratio of fuel energy used to electrical energy generated is called the *heat rate*. A heat rate for a power plant might be 10,000 British thermal units (Btu) per kilowatt-hour (kwh). In other words, 10,000 Btu of fuel (a gallon of gasoline contains 120,000 Btu) are burned to produce 1 kwh of electricity. Since a kwh represents 3,413 Btu, this plant operates at 34 percent efficiency. Plant heat rates are very important because they account for most of the utility's short-run variable costs. They are known as the utility's energy costs. Heat rates range from 9,600 Btu for the newest high-efficiency *combined cycle* plants up to 23,000 Btu per kilowatt-hour for old, quick-start jet turbines used only in high-demand emergencies.

Some Basic Electrical Concepts

The complex nature of the electrical system requires a fairly exact understanding of the three most important electrical elements: current, voltage and wattage.

Current is measured in *amperes* (amps for short). An amp represents the number of electrons flowing past a given point in a given period of time. Modern science allows us to quantify exactly how many electrons there are in an amp. With a current of exactly 1 amp, the number of electrons passing a given point in one second is 6,242,000,000,000,000,000 or, as physicists say, 6.242×10^{18}.

Direct current (DC) electrons move in only one direction. This is the type of one-way current that batteries deliver. In *alternating current* (AC), electrons reverse directions. This type of current is used in almost all of our electrical systems. The number of times electrons move back and forth determines the *frequency* of the current. The standard frequency in the United States is 60 cycles per second (60 *hertz* or Hz).

The load on a circuit is characterized by its resistance to the flow of electrons. This resistance is measured in *ohms*. The amount of resistance varies greatly depending on its composition, thickness, density, temperature and so forth. Insulators have very high resistance while conductors have very much lower resistance.

The pressure with which a generator pushes electrons through a circuit is called *voltage*. Like amps, volts can be DC or AC. The voltage of electricity that enters homes in the United States is 120 v. The higher the voltage the more easily the electricity can pass through materials. For example, a flashlight battery gives off about 1.5 v. A simple patch of rust on the switch or grease on the contacts will cut off the bulb. Anything less than a substantial thickness of material will not interrupt the flow of 120-v electricity.

Low voltages are safer to handle because they cannot penetrate the skin. Below 50 v there is little danger of a dangerous shock. A car battery operates at 12 volts direct current (vdc) and a heavy-duty truck battery at 24 vdc.

Higher voltages can be quite dangerous. At 132 kv, for example, sparks can jump through 2 inches of air. At 765 kv sparks can jump almost 2 feet through the air. The force of the electricity moving through high-voltage lines literally creates a crackling sound. The danger from these high voltages is one reason bulk power transmission towers are built so high off the ground.

As electrons move through a load (a light bulb, a motor) they do work. The rate at which work is done is known as *power* and is measured in *watts* (w) or *kilowatts* (kw). One kw equals 1,000 w, which in turn equals 1.34 horsepower (1 HP equals 0.75 kw). Incidentally, the basic unit of power, the watt, is named after James Watt, inventor of the steam engine.

The best way to think of electric power is "energy per hour," just as speed is defined in terms of distance per hour. If we multiply power (in watts) times the duration of power usage (in hours), we arrive at watt-hours or *kilowatt-hours* (kwh), the familiar measure of electrical energy. A 100-w bulb burning for 10 hours consumes 1 kwh of electrical energy. One kwh equals 3,413 Btu. Btu stands for British thermal unit and is the amount of heat required to raise the temperature of 1 pound of water 1 degree Fahrenheit. Incidentally, there are 4 Btu in a Calorie. Therefore, the 3,000 Calories the typical person consumes every day could, if converted into electrical energy at 100 percent efficiency, light a 100-w bulb for more than 40 hours!

Appliances are rated by manufacturers according to their power consumption. General Electric, for example, makes 25-w, 60-w and 100-w light bulbs. The power consumption of appliances is usually on the back plate. Sometimes the amps but not the watts are given, but watts are calculated as the product of volts times amps. Thus, by multiplying the household voltage (120) by the amp rating, you can derive the appliance's wattage. At household voltage, a 25-w bulb will draw a current of about one fifth of an amp (25w ÷ 120v), while the 100-w bulb will draw about four-fifths of an amp.

Just as appliances are rated according to power consumption, so are a utility's power plants rated according to the power they produce. Large central power plants are rated in millions of watts (megawatts). A 1,000-megawatt (Mw) plant running continuously for 24 hours will produce 24,000 megawatt-hours (Mwh) or 24 million kwh.

The power demand determines how many kilowatts of capacity a power plant must have on-line at any given time. The maximum amount of power demand experienced during a year tells the utility how much total capacity it must have available. Power demand and load are essentially the same thing.

Energy and power are two distinct elements. The timing of one's demand for energy is as important in planning an electrical system as the power or wattage of a particular appliance. The distinction between power and energy may be better illustrated by an example using the water consumption of a household sink and a garden hose. Each provides water at 5 gallons per minute (gpm). The sink holds 10 gallons and watering the garden requires 10 gallons as well. If you fill the sink and water the garden at the same time, the demand for water will be 10 gallons per minute (5 gpm + 5 gpm). The load imposed on the local reservoir, then, is 10 gpm. If the reservoir is designed to serve 100 customers, and all the customers use their faucet and hose simultaneously, the reservoir must be able to deliver water at 1,000 gpm. This in turn requires a powerful pump and a wide-diameter pipeline.

However, you could fill your sink first and water your garden afterward. The total amount of water needed remains the same at 20 gallons. But the instantaneous load imposed on the reservoir would be only 5 gpm. If all 100 customers use each outlet separately, the reservoir would need to pump only 500 gpm. The water-flow load is thus halved and thus a less-powerful pump and smaller-diameter pipeline can be used.

Water flow in gallons per minute is analogous to electricity flow rates in kilowatts. The reservoir is the counterpart of the electric power plant. The total water volume in gallons is comparable to the total energy consumption in kilowatt-hours. Both amounts are determined by the amount of time the faucets or appliances are operated.

Since power multiplied by time equals energy, the same amount of energy can be used but over a longer time period with less demand for power. For example, burning ten 100-w bulbs for one hour consumes 1 kwh of electrical energy. Burning one 100-w bulb for ten hours also uses 1 kwh. However, in the first case, the ten bulbs use ten times as much power as the single bulb. Although the energy (work done) is the same, the ten bulbs burning simultaneously require ten times as much power plant capacity. The dominant consideration in utility economics is demand for power, or *load*.

An understanding of the relationship of energy to power can aid a small power producer in understanding the dynamics of the Public Utility Regulatory Policies Act of 1978 (PURPA). Electricity produced at peak times of the day (when the load or demand is greatest) will receive a higher price than electricity produced during off-peak times (when the load or demand is much lower). Moreover, if the small power producer can generate sufficiently reliable power so that the utility can actually displace future generating capacity, the producer receives a bonus, a capacity in addition to the energy credit.

Load Characteristics

What effect does load have on a utility? To see how the use of household appliances might affect a utility, assume there is a group of 100 residential utility customers. For simplicity, also limit the appliance mix for each customer to the following:

10 electric lamps:	100 w each
4 portable space heaters:	1,322 w each
1 air conditioner:	1,566 w
1 hot water heater:	4,474 w

To analyze the power demand for this group you must know how a customer uses each appliance in relation to other appliances and how all customers combined use their appliances. Suppose all 100 customers simultaneously use all their appliances at their rated power demand. For each customer the load imposed would be 12.3 kw:

Electric lamps:	10×0.1 kw = 1.0 kw
Portable space heaters:	4×1.3 kw = 5.2 kw
Air conditioner:	1×1.6 kw = 1.6 kw
Hot water heater:	1×4.5 kw = 4.5 kw
	12.3 kw

The entire group of 100 customers would impose a peak load of 1,230 kw or 1.23 Mw. These 100 customers would need roughly 1.23 Mw of electric power generating capacity to meet their combined peak load. Generating plants are expensive. The typical plant starting up in 1980 cost about $1,000 per kilowatt to build. Therefore, this group of customers would have to pay roughly $1.23 million for the capacity needed. Each customer would have to pay $12,300 just to pay for the machines to generate enough power to be able to run their appliances all at once.

Fortunately, the situation is not this onerous. All the appliances don't run at the same time. The total system load is usually less than the sum of the individual loads. Appliances such as electric water heaters or refrigerators have intermittent demand characteristics. Water heater elements, for example, typically operate on a 25 percent *duty cycle*. That means the heating elements are on 25 percent of the time, and it can be said that the probability is that only 25 percent of all water heaters will be drawing power simultaneously. This is called a 25 percent *load diversity factor*. By applying the load diversity factor to the power consumption of a typical water heater, there is a diversified load of 0.25×4.5 kw = 1.125 kw.

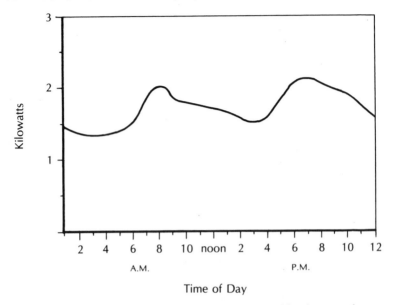

Figure 2–2:   This graph represents a typical residential load pattern for customers of the Pennsylvania Power and Light Company for the month of January 1981. This represents loads for the entire residential class. The load pattern for an individual household would demonstrate higher peaks and lower valleys. Notice peaks at breakfast time before people leave for work and at dinner time after they return from work.

Actually, load diversity will vary somewhat over the day. People will do the laundry or shower or wash dishes at different times. Demand for hot water doesn't vary greatly by season, but demand for space heating and lighting does. Since most work is done during the day, the peak load for most electric utilities occurs sometime between 12 noon and 9 P.M. The *base load,* or the minimum demand on the system, occurs sometime late at night. The diurnal or daily load curve of a utility as well as the seasonal load curves can be graphed. These curves will depend greatly on the climate, the saturation of major appliances and the composition of the consuming classes. Some utilities might sell most of their power to industries that operate 24 hours a day, 365 days a year. These would tend to have flat load curves on a daily or seasonal basis. Other utilities that have a large air-conditioning load are called *summer peaking,* and those with a large space-heating load are called *winter peaking.*

When off-peak demand and peak demand are similar, the system is said to have a *high load factor.* When there are sharp spikes in

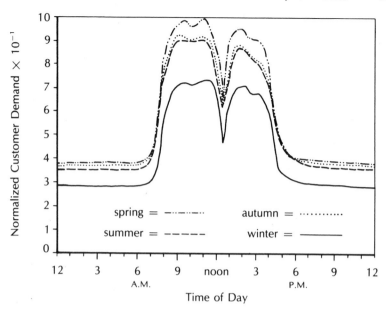

Figure 2–3:   This graph shows the seasonal average daily profile for the stone, clay and glass products industry. The maximum demand in kilowatts is 747.9. *Redrawn from R. A. Whisnant, C. B. Morrison, N. G. Staffa and R. D. Alberts,* Application Analysis and Photovoltaic System Conceptual Design for Service/Commercial/Institutional and Industrial Sectors, vol. 1 *(Albuquerque: Sandia Laboratories, 1979), p. 172.*

demand at brief intervals during the day or year and the rest of the time demand is low, the system is said to have a *low load factor*. The higher the utility's load factor, the more baseload capacity it will have relative to total capacity. In other words, the closer base load is to peak load, the more the utility can fill the total demand with efficient baseload power plants, which usually generate inexpensive electricity. This is because the total utility plant is being used more fully so that the cost of the plant itself is spread over more kilowatt-hours sold.

The more of the utility's load curve that can be met with baseload power, the lower the buyback rate for the small power producer. Baseload power, as will be seen, is the cheapest to operate.

Load curves for individual industries have their own special characteristics. A load curve for the textile industry would show that it runs on three shifts all year long. The vast amount of electrical energy is used to power its looms. It has a flat seasonal and daily profile. At 2 A.M. in December it needs about the same power as at 3 P.M. in the spring. Figure 2–3 illustrates the load profile of the stone, clay and glass

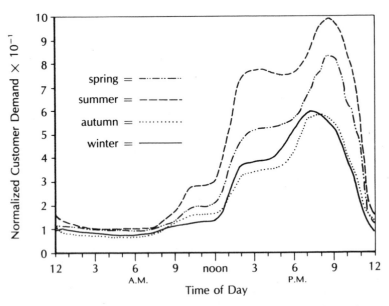

Figure 2–4:   This graph shows the seasonal average daily profile for the movie theater industry. The maximum demand in kilowatts is 25.72. *Re-drawn from R. A. Whisnant, C. B. Morrison, N. G. Staffa and R. D. Alberts,* Application Analysis and Photovoltaic System Conceptual Design for Service/Commercial/Institutional and Industrial Sectors, vol. 2 *(Albuquerque: Sandia Laboratories, 1979), p. 174.*

products industry. It demonstrates a one-shift operation, with greater production in the warmer months. The steep drop at noontime indicates a highly unionized labor force that gets an hour for lunch and a plant that basically shuts down during that time. A load curve for the personal services industry would show the demand for services begins in the morning from 7 to 8 A.M. and reaches its maximum at about noontime, steadily declining till 8 P.M. The peak summer demand for power in these businesses is about twice that of its demand in winter, autumn or spring. The industry pattern illustrated by figure 2–4 is obvious even if the name weren't given. The very low demand in the early morning hours that rises gradually as the businesses begin to open at about noon or 1 P.M. and then the steep rise at about 3 P.M., continu-ing to rise to the peak demand at 8 P.M. and then collapsing quickly at 11 P.M. clearly illustrates the entertainment, more specifically, the movie theater industry.

Filling the Load Curve

The utility fills the load curve by bringing into operation different types of power plants. It meets the demand from the bottom up. The *base load* represents constant demand or load that is always present. Since the utility is required to generate electricity as cheaply as possible, this constant portion of load is met with the most efficient capacity. The utility wants to use a plant with the lowest heat rate that uses the cheapest fuel. It is now known that the heat rate represents the amount of fuel burned to generate a kilowatt-hour of electricity. So the plant wanted is the one that burns the least amount of the lowest-cost fuel to generate a kilowatt-hour (see Appendix 4 for prices of electricity from various utilities).

Baseload plants are efficient but expensive. Today's baseload power plants cost between $800 and $3,000 per kilowatt. Fired by coal or uranium, or in some cases "heavy" or less refined oil, they have heat rates of about 10,000 Btu per kilowatt-hour. California, for example,

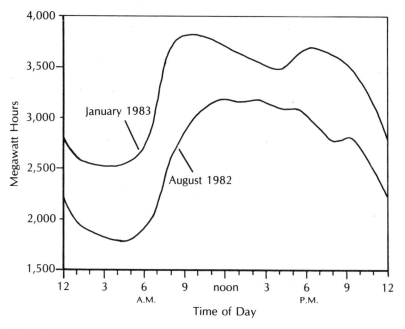

Figure 2–5: This graph shows average weekday load curves for January 1983 and August 1982 for the Pennsylvania Power and Light Company. *Redrawn from a graph provided by the Pennsylvania Power and Light Company.*

has strict environmental regulations that prohibit coal burning, so most of California's baseload plants are oil fired. Baseload plants also tend to be the newest plants. These plants operate 4,000 to 6,500 hours out of an 8,760-hour year.

Intermediate load is met with cycling units, which are usually fired by oil. These are less efficient than baseload units but cheaper to build. They have heat rates of about 11,500 Btu per kilowatt-hour. They tend to be middle-aged plants, that is, former baseload units relegated to less frequent service. Intermediate load plants operate about 1,000 to 4,000 hours a year.

Peak load is met by plants that are the cheapest to build and the most expensive to run. A typical peaking plant costs between $300 and $600 per kilowatt. Peaking plants are characterized by high heat rates, usually in the range of 13,000 to 16,000 Btu per kilowatt-hour, and the use of expensive fuels, such as oil and natural gas. Some very old oil-fired peaking plants might have heat rates as high as 25,000 Btu per kilowatt-hour. But because of their function, peaking plants usually operate only 50 to 1,000 hours a year.

Because of the higher capital cost of baseload plants, they are cost-effective only if they are run as much as possible. In this way they can spread the high capital cost over more kilowatt-hours of production, thereby lowering the average cost per kilowatt-hour. Baseload plants don't operate 100 percent of the time in part because of their forced outage and maintenance schedules.

Power Plant Performance

Each power plant facility has a name. Because of increased attention to electric rates and nuclear power, some of these names have become almost household words: Seabrook, Diablo Canyon, Bailey, Three Mile Island. The name refers to the entire complex, though each complex often contains two or three power plants, called units. Three Mile Island, for example, has three units. While Units I and II were shut down during the accident in 1979, Unit III continued to operate.

About 80 percent of our country's electricity comes from 900-odd steam plants. In 1980 steam plants generated about 58 percent of their electricity from coal, 12 percent from oil, 17 percent from natural gas and 13 percent from uranium. About half of our total electrical capacity comes from fewer than 300 power plants. About 12 percent is generated by 1,200 hydroelectric dams and 8 percent by 1,200 gas turbines (which run on average only 7 percent of the time). One percent comes from nearly 1,000 diesel generating engines located mainly in rural areas or used as small peaking plants.

Each power plant has its own unique operating characteristics. Large power plants are rated in millions of watts (Mw). An 800-Mw power plant, operating at nameplate capacity 100 percent of the time, would generate 8,760 hours times 800,000 kwh, or 7 billion kwh a year. However, no plant operates at full capacity nor all the time. The performance of the plant can be evaluated by comparing its energy production in kilowatt-hours with its nameplate rating in kilowatts. This relationship is known as the plant's *capacity factor*.

The capacity factor is a useful tool because it indicates how often the plant is generating electricity. The capacity factor includes all types of *downtime* (called outages). A failure in a major component may shut the plant down completely or partially. In the latter case the plant is said to be temporarily *derated*. Planned outages for maintenance may also be partial or complete and usually take place during off-peak periods.

Different types of plants have different outage characteristics. Nuclear plants tend to have longer outage periods than nonnuclear plants. For example, if Metropolitan Edison's Three Mile Island plant had been a coal plant instead of nuclear, the anxiously awaited "cold shutdown" would have happened in a matter of hours instead of weeks. In addition, repair crews would have had to worry only about temperature and noxious gases instead of radiation contamination.

Capacity factors directly affect the cost of electricity. For example, utility planners initially believed that even if nuclear plants were more expensive to install than coal plants, their low fuel costs would make nuclear-generated electricity cheaper than that derived from coal. But nuclear power is cheap only if the plants are running most of the time. Since nuclear plants include a higher proportion of fixed capital costs than any other type of power plant, an idle nuclear plant is a heavy financial burden on a utility. And since nuclear plants are relatively new, planners relied on the industry's estimates of 70 to 80 percent capacity factors to determine the cost per kilowatt-hour. But major studies conducted by economist Charles Komanoff in the middle and late 1970s concluded that the average nuclear plant delivered less than 60 percent of its nameplate rating during a typical year.[2] Accordingly, the cost of a kilowatt-hour of nuclear electricity is as much as 25 percent greater than previously assumed.

There is also a growing body of evidence supporting the view that there are significant diseconomies of scale. Komanoff demonstrated a clear relationship between scale and performance for both nuclear and coal plants. For each 100 Mw increase in size, the capacity factor fell off 3 percent for a nuclear plant and 2 percent for a coal plant. Robert Mauro of the American Public Power Association has remarked:

The disappointing availability record of many large units has diminished, if not entirely dissipated the theoretical savings expected from bigness. . . . Many small . . . electric utilities, which have been jeered at for operating "obsolete" plants with "tea-kettles" have had fewer problems in maintaining adequate power supply than some larger systems with modern large-scale units.[3]

Amory and Hunter Lovins point to one reason for the greater unreliability of large plants. "A five-hundred megawatt boiler has approximately ten times as many miles of tubing as a fifty-megawatt boiler, so 'a tenfold improvement in quality control is necessary to maintain an equivalent standard of availability for the larger unit.' A larger turbine has high blade-root stress, often forcing the designer to use exotic alloys with unexpected characteristics: highly skilled turbine designers in several advanced industrial nations have watched their turbines explode because the metal did not behave as hoped. . . . A more complex control system runs up against the discouraging mathematics of unreliability."[4]

One of the difficulties investigators have had in estimating the costs of electricity from various plants is the poor quality of plant-operating data. Up until 1979 data was collected by the Edison Electric Institute, and the reporting of data by member utilities was strictly voluntary. Now data collection responsibilities have been given to the National Electric Reliability Council, that arm of the utility industry charged with reliability planning. Many utility leaders are concerned with past inaccuracies and recognize the importance of these data for planning purposes. Also section 133 of the Public Utility Regulatory Policies Act (PURPA) requires large utilities to make public accurate cost data.

The capacity factor is a revealing statistic, but it has its limitations. It is most applicable to baseload plants because these are supposed to run full time. Since the capacity factor includes all types of plant derating and downtime, both planned and unplanned, it can distort some plant economics. Intentional curtailment of plant output, called *load following,* is supposed to occur only with cycling and peaking generating units. Utilities "follow" rises in the hourly load curve by throttling up the next-cheapest generating plant and likewise throttling back the most expensive as demand recedes. Load following is not supposed to occur with baseload plants. For optimum economic efficiency, baseload plants are supposed to be fully loaded at all times. They should be shut down only by component failure and for scheduled maintenance or refueling. Thus Komanoff initially found that coal plants have a relatively low capacity factor of 67 percent. But many coal plants are old and are used for load following. Almost 10 percent of the lost capacity factor for coal plants was due to load following. (Only 2 percent of the lost capacity factor for nuclear plants was for this reason.) If this discrepancy is taken into account, the capacity performance of coal plants rises to 75 percent.

Small power producers can expect to become expert at discussing capacity factors with utilities and public service commissions. Capacity factors serve as the basis for providing a capacity credit. Utilities will try to exaggerate their plants' capacity factors in order to reduce the amount they have to pay independent power producers. For example, some utilities will cite an 80 percent capacity factor for their own power plants and then require the independent producer to have at least that level of reliability to qualify for any capacity credit whatsoever. In this instance, a small power producer that can convince the public service commission that such a capacity factor is unreasonably high has a better chance of gaining a higher price for his or her electricity. In some states small power producers have persuaded regulatory commissions to require utilities to pay a proportional capacity credit. Thus if the small power producer achieves a 60 percent capacity factor but the utility standard is 80 percent, the capacity credit will be three-quarters the amount received if the independent producer had achieved parity.

System Load Variations and Plant Dispatch

Utilities with their own generating plants (as opposed to those that purchase all their power) have a central *dispatcher* whose job is to monitor system load and to ensure that the load is met at a minimum energy cost. In fact, the dispatcher has an unenviable task. The dispatcher (with the aid of sophisticated computers) must be familiar with the typical daily swings in system load and must anticipate these changes by making sure that adequate *spinning reserve* is on-line at any given time. Spinning reserve is a plant that is only partially loaded, ready to be "throttled up" to fuller output to accommodate greater system demand. Maintaining adequate spinning reserve makes it hard to run the system at the absolute minimum energy cost. For example, suppose a utility has one large coal plant (800 Mw), one 600-Mw, oil-fired cycling plant, and one 200-Mw peaker. Suppose further that demand at 8 A.M. is 750 Mw and is expected to rise to 1,000 Mw by 10 A.M. To meet load at the "absolute" minimum cost, the dispatcher would want to meet that 8 A.M. demand entirely with inexpensive, coal-fired generation. However, since the load will increase by 250 Mw within 2 hours, the oil-fired plant must be "spinning" in time to take on the additional load. Some large plants take 24 hours to get up and running. Smaller units can be called into service within hours from a "cold start."

Spinning reserves are made up mostly of partially loaded cycling plants. *The efficiency of plants varies directly with the fraction of the capacity that is being used.* A partially loaded plant has a higher heat rate and, therefore, high energy costs. Returning to the example, the dispatcher will probably not meet the 750-Mw load at 8 A.M. entirely with the coal plant. Instead, the load will be served by something like 700 Mw of coal and 50 Mw from the cycling unit. As the system load begins its daily ascent, the coal plant will follow the load up to its full capacity, or close to it, with the cycling plant resting for most of the rest of the day. Running the next plant at partial output in anticipation of a load swing is sometimes called *ramping*.

In the real world, things get even more complicated. Most utilities operate within power pools. Because of interchange agreements, utility dispatchers often end up shopping around among one another on the grid in search of power that can be bought more cheaply than it can be generated at the margin. *In other words, if the next kilowatt-hour can be imported more cheaply than it can be produced, then it will be bought rather than generated.*

This policy is reasonable and provides the basis for utilities to purchase power from small power producers. However, in practice, utilities are much more comfortable buying electricity from other utilities than from individual small producers. Independent producers will increasingly discover that utilities refuse to buy from them, or reduce the buyback rate because they are getting cheaper electricity from the interties. For example, in 1983 the heavy rains of 1982 and diminished electrical demand in the Pacific Northwest gave the system such a surplus of hydroelectric power that it literally had to dump water from its dams. This so-called dump power was sent through the Pacific Intertie to California at an extremely low rate, competing with small power producers who were counting on long-term contracts at much higher rates. Although the dump power was a temporary situation, it was sufficient to discourage some from entering the power production market.

Case Study: Boston Edison

Every year investor-owned electric utilities must file a report with the Federal Energy Regulatory Commission (FERC), the independent agency charged with regulating interstate sales of gas and electricity. The annual report, also known as FERC Form 1, contains a wealth of information about the operating characteristics of the utility's power plants.

Since the small power producer is going to negotiate a contract based on the avoided cost of power to the utility, he or she needs to know intimately the costs of operating the utility system. To obtain a copy of your utility's Form 1, contact the corporate finance department. If the company will not mail you one, contact your state utility regulatory commission. It has a Form 1 available for public inspection.

Form 1 contains financial statements, expense statements, and production and sales statements. Financial statements will probably be comprehensible only to trained accountants. The expense sheet lists just about every aspect of corporate operation, including political and social contributions, expenditures for regulatory proceedings, advertising, research and development, and current construction outlays. The production and sales section contains data on plant efficiency and availability.

Under the heading "Steam-Electric Generating Plant Statistics," Form 1 contains the following information about each facility.

1. Type of plant and years when first and last generating units were added
2. Total capacity in kilowatts
3. Net peak demand on plant
4. Hours connected to load
5. Net continuous plant capability
6. Net generation in kilowatt-hours
7. Cost of plant, total and per kilowatt
8. Production expense—total, fuel and per kilowatt-hour
9. Fuel—kind, quantity consumed, heat content, cost per unit, cost per million Btu burned, average cost per kilowatt-hour generated
10. Average plant heat rate

These figures apply only to each plant. They are not broken out by each unit of each plant.

The information on the following pages is from Boston Edison's Form 1 for 1981. Boston Edison's generating plants are oil-fired except for one nuclear plant, Pilgrim I. Adding up the nameplate ratings of all the oil-fired plants totals 2,079 Mw of capacity. The Pilgrim I nuclear station has a nameplate rating of 687 Mw. Therefore, Boston Edison's supply capability mix is 72 percent oil and 28 percent nuclear.

Boston Edison uses its nuclear plant more than its oil-fired plants. Form 1 shows that the generation output mix for 1981 was 68 percent

TABLE 2-1
Plant Capacity Factors and Energy Costs
for Boston Edison (1980)

PLANT	CAPACITY (Mw)	CAPACITY FACTOR (%)	ENERGY COST (¢/kwh)
Edgar GT*	28.4	0.2	11.22
Framingham GT*	42.6	0.2	10.26
L Street GT*	18.6	1.0	10.53
Mystic GT*	14.2	0.2	10.28
Mystic 200	468.7	32	5.96
Mystic Unit 7	617.0	51	4.84
New Boston	717.0	68.2	4.98
Pilgrim 1	687.0	57	0.4
West Medway GT*	135.1	1.7	4.61
W. F. Wyman Unit 4†	37.2	0.2	5.65

SOURCE: Information in this table is from Boston Edison's FERC Form 1 for 1981.
* GT means the plant is a gas turbine.
† Boston Edison owns 5.881 percent of this plant.

oil and 32 percent nuclear. That should come as no surprise. Pilgrim I's fuel cost in 1981 was 0.40¢ per kwh, or 4.0 mills. The oil cost at the time for the New Boston plant was about 12 times that or 5¢ per kilowatt-hour. Indeed, given the low operating cost, Boston Edison should want to operate its nuclear plant as much as possible. In 1978 its capacity factor was 72.7 percent, but in 1981 it dropped to 57 percent. The plant was only connected to load 65 percent of the year, indicating serious operating problems.

Table 2–1 gives the plant capacity and capacity factor and energy cost for each plant. The capacity factor can be estimated by multiplying the nameplate rating by the number of hours in a year, 8,760, and then dividing that into the total electricity actually generated by the plant during the year. For example, the New Boston plant is rated at 717 Mw. If it had operated 100 percent of the time, it would have had an output of 6,300 gigawatt-hours (gwh; 1 gwh equals 1 billion kwh). But it only produced 3,500 gwh for a capacity factor of 68 percent.

As seen, the merit order dispatching procedure should lead to the cheapest operating plant being used the most. There should be an inverse relationship between the capacity factor and energy cost. This proves to be true here. The very high cost gas turbines operate at less than 1 percent capacity factor. The only plant that appears to be an exception is the West Medway GT. It used gas as well as oil and with its relatively low operating cost, one might expect it to have had a higher capacity factor. This reveals that the plant was probably not kept from operating because of economy dispatching but because of forced outages.

Table 2–1 can be used to classify Boston Edison's plants according to dispatching merit order: base, intermediate and peak load. Pilgrim I, New Boston and Mystic Unit 7 could classify as baseload plants. Thus its baseload is in the neighborhood of 687 + 718 + 617 = 2,022 Mw.

Cycling or intermediate plants should demonstrate a capacity factor of 20 to 40 percent. Plants having capacity factors of less than 10 percent would be peaking facilities. However, the plants must show appropriately higher energy costs to fit these categories.

Another guidepost in determining plant type is to determine the hours connected to load. Form 1 provides a means to do this. Peakers usually operate fewer than 1,000 hours. Cycling plants operate 1,000 to 4,000 hours, and baseload plants operate over 4,000 hours.

The relative use of intermediate and peaking facilities will depend on two factors: (1) The shape of the utility's load curve, and (2) its access to power from neighboring utilities at lower cost than power available from its intermediate and peaking facilities. In general, the lower the cost and the greater the availability of electric power from the grid, the less use your utility will make of peaking and intermediate plants.

If short-run capacity needs can be met with purchased power at a price lower than the marginal operating cost, utilities will substitute purchased power for electricity they generate themselves. Usually purchased power does not displace baseload generation unless the base load is oil fired.

Mystic 200 at 469 Mw is Boston Edison's only candidate for an intermediate load plant. It was connected to load 98 percent of the time (8,604 hours) but with an average load of 150 Mw, it had a 32 percent capacity factor.

All other Boston Edison plants demonstrated capacity factors of less than 1 percent. These are exceptionally low and suggest the influence of purchased power availability or a flat load curve (high load factor). Form 1 helps here. The page titled "Electric Energy Account" shows that Boston Edison purchased about 10 percent of its electricity from the interchange. It also shows its load curve. The peak demand comes in the summer months. The peak hours in the summer are in midafternoon, while the winter secondary peak occurs in the early evening. Interestingly, Boston Edison in 1981 sold 10 percent more electricity during the month of January than it did in the month of July, even though its July hourly peak was 10 percent higher than its January peak.

The marginal costs for each load category can now be determined. Marginal or avoided costs will be explored in greater detail in the next chapter. To complete this exercise, take the highest cost of the plants in each load category, because, according to PURPA, the qualifying facility (QF) must be paid a price equal to the cost of the most expensive plant operating at the time. For base load in 1981 the cost was 4.98¢ per kilowatt-hour. For intermediate load electricity from the single Mystic 200 plant, the cost was 5.96¢ per kilowatt-hour. Boston Edison has several peaking plants. Including the Medway GT as a peaking plant, the cost ranges from 4.61¢ to 11.22¢ per kilowatt-hour. The average is 7.59¢ per kilowatt-hour.

Updated fuel statistics are available on a monthly basis from the U.S. Department of Energy's Energy Information Administration. The report is called *Cost and Quality of Fuels for Electric Utility Plants*. It shows fuel prices and Btu content for gas, oil and coal, as well as wood chips, coke and refuse-derived fuel for the third preceding month. In other words, the utility fuel prices listed are three months old.

According to PURPA, the small power producer must be paid the full avoided costs to the utility. PURPA will be discussed in detail in the next chapter. One of its primary tenets is that utilities pay the small power producer their full avoided costs. That is, the utility should pay the small power producer the price that it would cost the utility to generate an additional kilowatt-hour of electricity. Included in this cost are the line losses incurred when the utility delivers electricity over long distances. When a producer generates power in the distribution secondary, he or she avoids not only the fuel costs but the line losses as well. Form 1 indicates what the utility's line losses are. In 1981, 5.9 percent of the electricity generated by Boston Edison was lost in transmission and distribution. Thus the above prices should be multiplied by 1.059.

There are several reasons not to recompute the short-run energy costs of the nuclear plant, Pilgrim I. The main reason is that nuclear units are rarely "throttled back" for economy dispatch. The capital costs are too high for nuclear plants to follow load. Nuclear plants rarely carry the top of the utility's load for a significant portion of the time. The reader will see that, as a result, the Pilgrim I plant has a very low energy cost, but a very high capacity cost.

The following chapter will examine in detail the issue of nuclear power in PURPA calculations. The utility has, it will be seen, in effect substituted capital for energy. Higher fixed capital costs have been substituted for lower variable fuel costs. Therefore the $2,500 per kilowatt cost of nuclear is not, strictly speaking, the capacity cost of nuclear power. A portion of it could be viewed as the energy cost. A kilowatt of capacity for a new combined cycle peaking plant is $300 per kilowatt. The difference between the nuclear capital cost and the conventional capacity cost is known as the *energy related capital cost.* It reflects the long-run marginal energy costs of nuclear power. As will be explored in more detail later on, the utility's marginal energy costs are not simply its fuel costs in the long term. And while capacity costs are composed entirely of capital costs, not all capital costs are capacity costs.

It is worthwhile to explore the relationship of Boston Edison to its regional power pool. It has an interchange agreement with the power pool. For the pool as a whole it makes economic sense for several utilities to coordinate output under a common merit order dispatch. For example, one utility may peak at a different time than its neighbor. Instead of each utility individually meeting system load with expensive peaking plants, they could together use idle cycling or baseload generating units to serve each other's load partially on a reciprocal basis.

The extent to which utilities coordinate output in this way tends to vary according to how significantly their individual loads differ and the amount and types of generating capacity they have on-line. In the most extreme and most centralized case, one utility owns a number of subsidiary utilities, each with its own generating mix. All are coordinated by a "common economy dispatch" as if there were just one utility service area. Such is the case with the New England Electric System, the Southern Company and the American Electric Power Company. Normally, however, utilities alternately buy and sell from one another under a wide variety of wholesale purchase agreements approved by the FERC. For example, Public Service Electric and Gas (PSE&G) is a member of the Pennsylvania-New Jersey-Maryland power pool (PJM). It serves member utilities in Delaware, Virginia and Washington, D.C., as well. The transmission system delivers emergency power from sources in Canada, the Midwest, New York and elsewhere. In 1981 over 40 percent of PSE&G's energy requirements were satisfied from sources outside of New Jersey. The same year it sold 270,000 Mw hours of electric energy to other systems, which meant it was buying more than it was selling. PSE&G has 14 transmission and distribution interconnections that link its service territory with the other New Jersey utilities and 13 other transmission and distribution interconnections to utilities in New York, Pennsylvania and Delaware.

The exchange position of any utility will depend on regional circumstances. Regional demand and supply conditions are fluid. The following describes the interstate power market as of 1980.

Some systems are net exporters of electricity, such as those in the Southwest. Other systems, such as those in New England or California, are net importers. However, at any given time any system can be importing or exporting. And within regions we will find individual utilities that import most of their electricity even though the region is a net exporter. Once again, the small power producer must know the position of the local utility in the broad regional power pool to negotiate a contract intelligently.

New England, especially Massachusetts, depends heavily on oil, most of which is imported. Considerable nuclear capacity is currently under construction or seeking licensing. Boston Edison's abnormally low peaking plant capacity factors suggest that it relies on economy interchange agreements to cover its peak loads. New York State has been a big supplier of electric energy and capacity to the New England region, largely via the Power Authority of the State of New York. While New York City's Con Ed is relatively oil-dependent, the state has a good mix of hydro, nuclear and coal generation. The region is currently a big exporter with relatively little power imports.

The Mid-Atlantic/Midwest utilities are primarily exporters with a predominance of coal and nuclear generation. The giant American Electric Power Company, a holding company in the Midwest, is an especially big interchange supplier because of its access to a large capacity of reliable coal generation.

Southeastern United States is dominated by another holding company, Southern Companies, Inc. Nuclear is a heavy contributor in the region.

The South-Central region is dominated by the Tennessee Valley Authority (TVA), which is the federal bulk power supplier to many public and private distributors. It relies on hydro, nuclear and coal, although gas represents a significant part of the generation fuel outside of TVA in this region.

The far Western region, specifically California, has become a large interstate buyer recently, because the state Public Utilities Commission (PUC) wants to decrease the use of the state's preponderance of oil-fired plants and restricts the construction of coal or nuclear plants. While California's unique water pumping needs have been met from the Pacific Northwest via the federal Bonneville Power Administration

(BPA), purchases of all types from the Southwest region (Arizona, New Mexico, Utah and Colorado) are increasingly important.

The Pacific Northwest has long been a bulk power exporter because of its enormous hydroelectric capacity and the fact that its peak output coincides with southern California's peak demand. However, this region will reach the limits of its hydro capacity by the mid 1980s. Coal and nuclear will play an increasing role in its supply mix after that time.

System Reliability

System reliability is an important consideration in utility planning. *Reliability* is the predetermined acceptable rate of failure of a utility's entire system (when demand exceeds available supply). The utility standard is that such a situation should occur only one day in ten years. This is a very high quality service, and one that costs a great deal to maintain. All utilities install a certain amount of *reserve margin capacity*. As load grows and capacity remains fixed, the reserve margin shrinks, and the loss of load probability (LOLP) increases (reliability deteriorates). Conversely, if load were fixed and capacity were added, LOLP would fall (reliability improves). Reliability planning is most correctly applied to a *power pool*—a group of neighboring utilities—rather than to a single utility in isolation.

Recommended reserve margins have increased in the last 15 years. The reserve margin is the amount of capacity that's held in reserve, over and above the capacity to meet the peak demand. The recommended amount of reserve capacity has increased from 15 percent of peak demand in 1960 to 25 percent today. Still many utilities in the Midwest and Mid-Atlantic regions have reserve margins above 30 or even nearer 50 percent. Utilities in New York State have on average half of their entire system capacity standing idle even at peak periods of the year. This is taking place even while utilities learn that the vast majority of power disruptions occur because of failures in transmission and distribution facilities and not because of generating capacity shortages.

System reliability is under scrutiny because some customers who don't require a very high reliability factor may be paying the cost of maintaining the extremely high utility reliability standard. What exactly are the consequences if a utility lets its LOLP rise above one day in ten years? Another way of posing the same basic question is to ask what is the cost of underinvestment in capacity. Utilities would have the public believe that the cost is enormous—that a power shortage would produce calamity. The truth is that a temporary capacity deficit does not inevitably lead to a system collapse. Peak demand can be cut by reducing unnecessary power consumption and by automatically cutting off or cycling certain appliances (cutting off air conditioners 15 minutes on the hour in rotating fashion, for example).

The issue of reliability could have some important ramifications for small power producers. If customers decide to choose lower reliability (that is, a higher incidence of outages) in return for lower rates or internal storage or backup systems, the rates paid small power producers will change.

The costs of inadequate load-carrying capacity are different for different customer classes. A low-income residential community might not require the type of total system reliability standard for which utilities now strive. For them the only end use considered essential during the system peak might be food storage (refrigeration). On the other hand, an expensive hotel district might demand sufficient electrical capacity to meet all of its guests' needs and desires all of the time (especially for air conditioning). The residents of that district might accept a LOLP of one day in ten years. An industrial enterprise might need very high reliability because of the economic loss due to a plant shutdown even for a few hours. In fact, industries such as these often have an internal backup capacity. That is, they require even a higher degree of reliability than the utility can now guarantee. Hospitals also cannot afford a power blackout at any time. They are legally required to have back-up systems available in case of power outages.

The issue of reliability will play an increasingly important role as communities begin to install their own generating capacity. For example, a community might contract for a certain amount of firm service to supply what the members view as essential end uses. The rest of the desired service will come from a mix of wind, photovoltaics and cogeneration—electricity produced by and for the neighborhood itself with the surplus fed back into the grid for resale to other customers. These sales would be credited against the cost of the neighborhood's guaranteed minimum supply from the utility. Lowering the reliability requirements leads to cheaper power costs. Reliability refers to service quality with respect to system peak. An agreement to provide firm power off-peak carries with it no capacity costs, because by definition off-peak periods are those during which the utility has an abundance of unused generating capacity.

Residences do not yet have the ability to contract as proposed by this hypothetical community. But commercial and industrial enterprises now have the ability in most states to contract for "interruptible" power. The utility charges them less in return for being able to turn off their power at certain times of the year. The system saves money that would have had to be spent in adding new capacity, and the commercial customer shares in these savings.

Utility Planning

These days, a utility's plan to build a new power plant is a billion-dollar effort based on uncertain information about the future. Large generating stations may take 2 or 3 years for licensing and an additional 8 to 9 years for construction. If these plants are to come on-line in timely fashion, the utility must take action over a decade in advance of the actual need for electric power. To anticipate the amount of new capacity that should be initiated, the utility prepares 10- to 12-year forecasts of the demand for power.

Historically, electric demand followed a rather smooth growth trend, growing at an exponential rate of 6 to 7 percent a year. It doubled every ten years. One of the reasons for this fast growth was the decline in the price of electricity. Fuel prices dropped, and it was cheaper to build new power plants than to operate older, existing plants. Therefore, if the utility tended to overestimate future demand and build too many new plants, it could simply retire the older, more expensive ones somewhat sooner without imposing a large price penalty on the consumer.

Beginning in 1970, fuel and electricity prices began to rise. At first utilities and their regulators continued to use the old rules of thumb to predict future demand. As late as 1973, commissioners and utility staff maintained that, even if electric prices were to rise, the effect on customer demand would be negligible. They argued that the cost of power still represented such a nominal part of a business or household budget that higher prices would not change demand.

They were wrong. Mathematician Edward Kahn, of the Lawrence Berkeley Laboratory, examined the forecasts of the Edison Electric Institute, the industry's trade association, from the late 1960s through the late 1970s. Writing in 1979, he concluded "that forecasts are almost always too high and that accuracy is poor. Only once during the last eleven years has the industry underpredicted, and this was only by a tenth of a percent. More important, the consistent pattern of overestimates became much more exaggerated after 1974." Forecasts that were on average 2.1 percent too high between 1968 and 1972 were on average 5.1 percent too high after 1974. Kahn added, "The average forecast error is larger than the average annual growth rate over the last five years."[5]

The financial penalty attached to incorrect forecasts was exacerbated by the rising cost of power plants. Plants built since 1970 have proved to be substantially more expensive than their predecessors. Poor forecasting translates directly into higher rates. According to a 1979 article by *Public Utilities Fortnightly,* a utility trade magazine, the industry's reserve margin as of year-end 1978 was 38 percent. If one willingly accepts the industry's standard reserve margin of 20 percent, there is still an excess capacity of 18 percent. The investment (by investor-owned utilities) in this excess capacity exceeded the total of all new plant additions in the preceding four years, representing a total investment of $65 billion! Given the average cost of capital for utilities of 15 percent, this can be converted into an annual revenue burden for the entire industry of $10 billion, or 10 to 15 percent of these investor-owned utilities' total revenue requirement!

The revolution in the cost of power and the nature of demand changes led regulatory commissions to adopt a more aggressive role. Until the mid-1970s, regulatory commissions did little more than divide up the surplus money generated by new power plants and increasing sales. But as rates began to escalate and utility-sponsored demand forecasts proved inaccurate time and again, the regulatory commissions began to assume a role in overseeing investment strategies. Aided by state legislatures that gave them this increased authority, commissions began to examine the resource plan of utilities. In an increasing number of cases, commissioners recommended delay or cancellation of proposed power plants in the light of softening demand.

Increasingly they forced utilities to justify new plant expansions or specifically refused to allow rate increases to pay for additional capacity. In several states, encouraged or directed by state legislatures, the public service commissions required utilities to encourage energy conservation. California and Minnesota, for example, required utilities to become bankers and finance energy conservation. Wherever the utility could save a kilowatt-hour for less than it could generate an additional kilowatt-hour, the public service commissions required it to do so.

Some utilities, such as the New England Electric System, needed relatively little persuasion to move in the new direction. Their executives realized that conservation improved their internal balance sheet, improved the value of their stock and allowed them to borrow money at lower costs. Most utilities refused to be so farsighted. Commonwealth Edison and Con Ed were among the most reluctant to encourage conservation. Although they argued that with an excess system capacity conservation could do no more than raise the rates, they also argued that they would need additional capacity within the decade. Regulatory commissions increasingly found such arguments self-serving and internally inconsistent. The federal government added its authority to encourage this change.

In 1978 the U.S. Congress enacted two pieces of legislation intended to redefine the role of utilities. The National Energy Conservation and Production Act required utilities to provide energy audits for residential customers and to provide assistance in encouraging energy conservation programs. PURPA required state regulatory commissions and large unregulated utilities to collect and examine data that would exactly identify the costs of serving various customer classes. PURPA also required an examination of rate structures to identify how they might be changed to encourage investments in cost-effective energy conservation or renewable energy measures. Many states required utilities actually to finance energy conservation or solar technologies.

Utilities and regulatory commissions also improved their forecasting models, but this alone could not eliminate planning uncertainties. Differences in demand forecasts are unavoidable. Depending on the assumptions that are employed, a variety of different forecasts can be obtained. Andrew Ford and Irving Yabroff, two utility experts at the Los Alamos Scientific Laboratory, conclude, "The development of more sophisticated forecasting methods will not eliminate the underlying uncertainty that will accompany every forecast of the future demand for electrical power."[6]

One way to reduce uncertainty is to encourage smaller power plants to come on-line more rapidly. As Kahn says, "The extended planning horizon required for large projects imposes a serious risk of excess capacity. Shortening the lead time for supply reduces exposure to this risk."[7]

Looking at it from the insurance company's perspective, small power plants reduce the risk premium. They can more easily match changes in demand with changes in supply. Rather than making 12-year forecasts, the utility need only accurately forecast demand for 3 to 5 years.

Another advantage to small-scale systems was that they increased the overall system reliability. Systems with a diversity of small power plants tend to have smaller forced outage rates than those dependent on one or two large plants. Moreover, the reserve margins can be lower. One 1,000-Mw plant requires an equally large one in reserve, but ten 100-Mw plants require far less reserve capacity to guarantee the same reliability.

Rate Schedules and Marginal Cost Pricing

In 1975 *Business Week* reported, "Without some fundamental changes, the situation will soon mount to crisis proportions. Rate reform seems inevitable. Big users will have to pay more per kilowatt-hour rather than less, and customers who want power during peak demand periods will have to pay a premium for this privilege."[8] The problem with the traditional rate structure was that it assumed a declining marginal cost of power, so electricity was priced on the basis of past, rather than future, investments. The price paid for electricity today is based on the cost of all the transmission and distribution lines and power plants already built. Much of this investment occurred decades before, when power plants and transmission lines were much less expensive. In fact, much of this investment has already been written off through depreciation.

This investment represents a utility's rate base. State regulatory commissions allow utilities certain rates of return on their rate bases. The return varies depending on the state and the utility. Today it ranges from 11 to 17 percent. To this cost is added the cost of fuel, and the total is divided by the number of kilowatt-hours purchased. That gives the average retail cost per kilowatt-hour, which forms the basis for rates.

When the marginal cost of energy began to increase in 1970, the electric utility ratemaking structures no longer gave customers accurate price signals. New power plants cost considerably more than old ones. Yet customers have continued to pay prices that encouraged more consumption. The greatest discrepancy between old and new is still taking place in the Pacific Northwest. Most of its power comes from hydroelectric facilities built a half century ago. The investments in these facilities have long been repaid. Free hydropower "fuel" and low operating and maintenance costs combine to give the Pacific Northwest the lowest electric rates in the nation—an average of 2.5¢ per kilowatt-hour in 1982.

But with its large-scale hydroelectric capacity fully taken up by demand, the region has had to meet additional demand with coal or nuclear steam generating plants. New electricity from these plants can cost the utility 7¢ to 15¢ per kilowatt-hour. But when a new coal or nuclear plant begins operating, customers don't "see" that price. The new, high-cost electricity is blended in with the much larger quantity of very low-cost, old electricity. Rates creep upwards. Customers don't pay 7¢ to 15¢ but instead pay 2.8¢ or 3¢.

The result of this pricing policy is that the customer has no incentive to save electricity even though by doing so the entire region could save money by avoiding the need for new power plants. By installing attic insulation, for example, a customer might be able to save a kilowatt-hour of electricity for 3¢ (per kwh saved). To the utility, a kilowatt-hour saved is equivalent to a kilowatt-hour generated. The customer could thus "generate" a kilowatt-hour at less than half the cost of even the cheapest new steam generating plant. But the customer will not undertake such conservation, because the investment, while cost-effective from the utility's vantage point, is not cost-effective from the homeowner's perspective. Electricity is just too cheap for the homeowner to worry about conserving it. Meanwhile, huge rate increases wait just around the corner as the utility builds very expensive new power plants to meet demand that is increasing in large part because electricity prices are so low.

A utility's rate structure is described in its *tariffs*. Each class of customer has a different tariff, and there can be several tariffs for each category. Separate tariffs can exist for small businesses and large businesses, for residences with electric space heating and for those without, for street lighting and so forth.

Your monthly bill usually consists of three components: a customer charge, an energy charge and a demand charge. The *customer charge* is a flat amount that is paid whether you use electricity or not. This charge is supposed to represent the costs of administrative overhead, meter reading and billing associated with that customer class.

The customer pays a minimum *energy charge* even if the consumption that month were only 1 kwh. Thus the first 50 kwh consumed can cost up to 20¢ per kilowatt-hour. The energy charge can be flat, meaning that no matter how much one consumes, the charge per kilowatt-hour remains constant. Or the energy charge can be divided into tiers, with each block of energy usage charged at a different rate. The practice used to be that the more one used, the less one paid per kilowatt-hour. This is changing rapidly. Now flat rates and even inverted rates are increasingly common.

The third component of the monthly bill is the *demand charge*. Residences do not pay demand charges. These are charges based on the highest amount of power consumed in any small time interval during a month. Often the separate demand meter has a ratchet on the register. The highest demand reached for any 15-minute period is the demand for billing purposes for the entire month. Many utilities carry over the ratchet effect so that one pays in succeeding months the highest demand reached during that quarter or that year. The utility argues that it must have capacity available all year to meet the highest momentary demand and therefore should charge all year for that one 15-minute spurt.

For examples, examine three different tariffs from Green Mountain Power Corporation. For the Residential Rate 01, the customer charge is $5.50, an amount that a customer pays even if no electricity is used that month. Energy charges vary by season. The two tiers of energy charges represent a sharply inverted pricing structure, especially for the peak season, for that is when one is paying for capacity that is in excess (unused) during the off-peak months. (This utility's peak occurs during the five winter months.) This tariff is unusual in that it gives Green Mountain the right to install load management equipment for electric water heaters and storage space heaters. It can send a radio signal to turn off power to the storage systems at peak times essentially to limit peak loads.

Green Mountain's General Rate 06 applies to nonresidential applications where the demand is less than 500 kw during the peak months. A commercial customer pays a slightly higher minimum charge than the residential customer. The flat-rate structure of this tariff varies dramatically by season. The commercial business also pays a demand charge of 50¢ per kilowatt of connected load. Those with connected loads greater than 40 kw pay $1 for each kilowatt of measured demand. An interesting provision is that 1 HP is taken as 1 kw, even though it actually translates into 0.75 kw. Load control is also permitted here as in the first tariff.

Green Mountain's Power Rate 14 tariff is available to all customers except for use as supplemental or standby power. It is characterized by a flat demand charge and flat energy charge. The demand charge is almost twice as high in the peak as in the off-peak season, and the energy charge drops by one-third in the off season. This tariff illustrates ratcheting carried over into later months. The demand charge for any given month is based on a measured demand not less than 50 percent of the highest 15-minute peak occurring during the preceding 11 months.

This tariff also contains a provision for power factor adjustments, an efficiency factor that will be discussed in greater detail in the interconnection chapter. Basically, the lower the power factor the more power the utility has to supply to meet a given load. Here the demand charge is increased by 1 percent for each 1 percent by which the average monthly power factor lags below 85 percent.

The Power Rate 14 tariff also gives the customer a credit for avoided transformer losses if the power delivered is measured on the primary side of the step-down transformer. The customer charges are discounted by 2.5 percent. If the customer uses 2.3-kv electricity directly or furnishes the transformer, it receives a credit of 24.1¢ per kilowatt of demand used for billing purposes.

Rate structures are changing rapidly. Title I of PURPA encourages utilities and regulatory commissions to restructure rate schedules to encourage cost-effective conservation and small-scale power facilities. The review of existing rate structures is mandatory. But the revision of the rate schedules is voluntary. Most states are moving toward *marginal cost pricing,* which revises rates so that the customer pays the genuine cost of his or her demand on the electric system. This means introducing flat or even inverted rate structures in which the customer pays more for the last unit of electricity consumed than for the first. It also means introducing *time-of-day pricing,* where the customer is charged more for electricity consumed during the peak hours of the day or year than for those purchased during the off-peak hours. These new rate structures encourage customers to invest in conservation and also to level or flatten their load curves. Using automated equipment, some commercial enterprises have found that their individual load factors (base load divided by peak load) can thus rise from 50 percent or less up to 95 or even 99 percent. A 50 percent load factor means that the peak load is twice that of the base load. The firm saves by reducing its demand charges from the utility. The rest of the utility's customers and stockholders save by sharing in the savings gained from not having to borrow money for new power plants.

This chapter has described the basic functioning of the present electrical generation and transmission system, but the coming years should see a massive structural change in this system. Millions of additional generating facilities will be added to the existing 4,000 power plants. Compared to conventional plants, each of these new additions will generate only an infinitesimal quantity of electricity, but their combined capacity will be substantial.

The new era began in 1970 when for the first time in 100 years the marginal cost of energy increased. That event changed the nature of our electric pricing system. When marginal costs are declining, it is economical and efficient to encourage greater usage. The more demand, the lower the costs. But when marginal costs are rising, an opposite dynamic occurs. It becomes more efficient to discourage demand and encourage conservation. It also becomes more efficient to promote cost-effective small power production, especially of renewable-based technologies that are immune from future fuel price increases.

For the homeowner, apartment dweller and businessperson, the change was formalized in 1978 with the passage by Congress of the Public Utility Regulatory Policies Act. Along with regulations already discussed, that act made the development of decentralized power generation a national policy and made illegal any efforts by utilities or state regulatory commissions to obstruct that development. Chapter 3 provides further explanation of that act and its consequences.

CHAPTER 3
PURPA

In 1978 a flurry of energy-related legislation was enacted. Five individual pieces of legislation comprised the National Energy Act. One of those was the Public Utility Regulatory Policies Act of 1978 (PURPA). It was a controversial act, but almost all of the controversy surrounded Title I, specifically Section 133, of PURPA. This section of PURPA required large utilities to make public information showing their cost of serving different customer classes at different time periods. Every public service commission was required to hold hearings to evaluate specific rate changes for their impact on conservation. Section 133 requires commissions to examine time-of-day rates, seasonal rates, lifeline rates and inverted rates (where additional consumption is priced higher than initial consumption).

Title II, on the other hand, sneaked into the legislation almost unnoticed. Although many people now take credit for having authored this section, few people at the time were aware of its potential impact. Title II affects all public and private utilities of any size. It is mandatory. Thus its impact is much more substantial than that of Title I. Yet until the Federal Energy Regulatory Commission (FERC) issued its preliminary regulations to implement this section of PURPA, little notice was paid to its profound implications. When the FERC transformed the general goals of the legislation into hard-hitting regulations, the nation became aware of the act's dramatic implications.

PURPA Title II, Sections 201 and 210, perhaps more than any other piece of federal legislation, made energy efficiency a matter of national policy and opened the way for small power plants.

Prior to the enactment of PURPA, a cogenerator or small power producer had three problems. First, utilities would often refuse to buy electricity from these producers. A utility might allow the small producer to send electricity to it but it wouldn't pay for it. Or if a utility did pay, the price was often unfairly low. Second, utilities discouraged small power production by charging extremely high rates for back up or supplementary power. Third, selling electricity could subject the small power producer to complex and expensive state and federal utility regulation.

PURPA was a legislative attempt to overcome these three primary obstacles. The act directed the FERC, formerly the Federal Power Commission, to establish regulations to implement the law. Final rules implementing Sections 201 and 210 of Title II of PURPA were issued on 25 February and 20 March 1980.

These regulations aggressively promoted the Congressional goals. They ended the century-old monopoly utilities had held over electric power sales. After March 1981 all electric utilities in the nation, including rural electric cooperatives, municipal utilities, investor-owned utilities and federal power agencies were required to buy energy and capacity from qualifying small power producers using renewable resources or cogeneration. The renewable resources include wind power, hydropower, photovoltaics, wood and other biomass and even garbage. With these fuels a utility's customers can now become its suppliers as well.

PURPA sets forth the conditions for becoming a qualifying facility (QF). It describes the various ways the QF can sell its electricity to the utility and the basis upon which the price the utility pays must be calculated. It sets forth the data the utilities must provide to potential QFs on their avoided costs. QFs not only qualify to sell electricity to utilities (and buy supplementary or back-up power at nondiscriminatory rates) but are also exempted from state and federal utility regulations.

QFs can interconnect to a utility's grid system in two ways. They are entitled to use their power on-site and sell only surpluses to the utility. In some cases state regulatory commissions have let very small facilities literally run their regular watt-hour meter backward when power is sent to the utility. This is called *net billing*.

QFs can also choose to sell all the power they generate to the utility and buy all the power they consume from the utility. This procedure, called *simultaneous purchase and sale*, requires at least two meters.

The distinguishing factor between these two arrangements is the number of connections to the utility's system. Allowing simultaneous purchase and sale permits the QF to continue to be a regular customer while also selling electricity. Utilities have argued that a building that produces power sometimes and buys power at other times has a unique load configuration and should be charged higher prices for back-up or supplementary power. But a customer with two connections to the grid system in reality continues to buy all of its power just like a nongenerating customer. Thus, there is no cause for treating a generating customer differently from a nongenerating customer with respect to charges for back-up power. Indeed, in its preamble to the regulations, the FERC specifically states that a QF involved in simultaneous purchase and sale contracts has no need for back-up power.

To encourage investors, the FERC requires utilities to pay QFs a price for their electricity equal to the price the utility would otherwise have to pay to generate or to buy an additional kilowatt-hour or unit of added capacity. The FERC calls these the utility's *avoided costs*. Buyback rates constitute the most controversial and important part of PURPA, and much of this chapter is devoted to a full discussion of what avoided costs really are.

PURPA requires utilities to provide back-up power to QFs at non-discriminatory rates. The QF cannot be charged any higher rate for back-up, standby or maintenance power than any other customer who doesn't generate power. The only exception is if the utility can prove that the cost of providing this service to the QF is higher than the cost of providing the equivalent service to a regular customer. This regulation is valuable not only to those who want to sell power. It also can prove immensely useful to those who produce only for their own use but want to use the utility as a backup. The homeowner who produces a portion of his or her own power cannot be treated differently from the next-door neighbor who is a regular utility customer.

Finally, PURPA exempts QFs from most federal and state financial, rate and organizational regulations that apply to electric utilities, although the act does not exempt them from federal, state or local environmental regulations.

PURPA has demonstrated the federal government's commitment to the generation of cost-effective electricity from renewable resources and high-efficiency cogeneration plants. During the law's passage and implementation, the following potential benefits of small-scale power production were noted:

1. Cogeneration uses fuels more efficiently than when thermal energy and electrical generation are accomplished separately.
2. Alternate energy sources diversify the utility's resource plan. By minimizing its dependence on any single source of generation, the system becomes more resilient, that is, less vulnerable to external supply disruptions (lapses in availability).
3. Electrical generation from biomass (including wood waste and refuse) or wind, water and sunlight makes the nation less dependent on foreign nations for fuels.
4. The development of many small power plants improves the total electrical system's reliability. The probability of many small plants failing simultaneously is far less than the proba-

bility of one or a few large central station plants suffering a forced outage.

5. The lead time required for construction of small facilities is several years shorter than that for a large central station power plant. Shorter lead times lessen the risks associated with construction cost overruns, rising interest rates and changing demand. Small facilities have also simplified licensing and siting requirements, as do cogeneration systems located at existing industrial facilities.

6. Many small power plants lessen the reserve margin that the total system requires. Large plants need equally large back-up plants. Because of their higher overall reliability, many small plants need proportionately smaller back-up plants for reserve capacity.

7. Independent power producers raise their own capital. The utility, therefore, does not have to raise the capital to construct the facility. The facility is not included in the utility's rate base and thus lessens the possibility of poor bond ratings and higher interest, leading to higher capital costs. The ratepayer does not have to bear the cost of unscheduled plant outages.

8. Electrical generation from many renewable resources promises to be more environmentally benign than generation from fossil fuels or nuclear power.

Sections 201 and 210 of PURPA apply to every utility in the nation—even those that do not generate electricity themselves. For regulated utilities, the state public service commissions have the responsibility for implementing the act. In those states in which the regulatory commissions have no authority over rural electric cooperatives or municipal utilities, the rural electric board and the city council must exercise this responsibility. State utility regulatory commissions and unregulated utilities were required by 20 March 1981 to have adopted rates or procedures for determining rates, or to have at least announced that they would entertain petitions from QFs for interconnection and rate determinations.

As it has turned out, several state legislatures have enacted mini-PURPA laws, putting these states on record as encouraging small power production. The first to do so was New Hampshire with its Limited Electrical Energy Production Act of 1979. Indiana, Minnesota, Oregon, New York, North Carolina and Montana followed New Hampshire's lead. These state laws became important when several law suits put key PURPA regulations in doubt. They gave a legal basis for public utility commissions or city councils to continue implementing the law.

Congress and the FERC gave states and unregulated utilities broad authority to provide even greater incentives to small power producers than did PURPA, which is essentially a minimum guideline. Thus, for example, North Carolina specifically required long-term contracts. Oregon required distribution utilities to pay a much higher price for QF energy than they would have had to pay under PURPA guidelines.

Most regulatory commissions and unregulated utilities have been slow to implement PURPA, in some cases because of reluctance or even hostility to the new law. But in most instances the delay was simply a matter of their entering unfamiliar territory. Not only was the field a new one, but the principal actors were given ground-breaking responsibilities. Attorney Peter Brown is a specialist in small-scale

---

TABLE 3–1

Required Data for Utilities' Avoided Cost of Energy

UTILITY SIZE	LEVEL OF PURCHASE FROM QUALIFYING FACILITY(IES)	UNIT OF COST	CONSUMP- TION PERIODS	YEARS REPORTED
Each utility system with a peak demand of 1,000 Mw or greater	Blocks of not more than 100 Mw	¢/kwh	Daily peak off-peak Seasonal peak off-peak	Current calendar yr Each of the next 5 yrs
Each utility system with a peak demand of less than 1,000 Mw	Blocks of not more than 10% of system peak demand	¢/kwh	Daily peak off-peak Seasonal peak off-peak	Current calendar yr Each of the next 5 yrs

NOTES: From Randi Lornell, "A PURPA Primer," *Solar Law Reporter*, May/June 1981, p. 45.
The data in this table is required by PURPA Section 210 of all utilities with sales of at least 500 million kwh annually. Utilities with sales greater than 1 billion kwh annually must have reported these data by 1 November 1980. Utilities with annual sales of between 500 million kwh and 1 billion kwh must have reported by 30 June 1982.

TABLE 3–2
Required Data for Utilities' Plans and Cost

CAPACITY TYPE	AMOUNT	COST OF COMPLE- TION	ENERGY COSTS	FREQUENCY OF COST CALCU- LATIONS
Capacity additions	By individual generating units	$/kwh	¢/kwh	Each yr for the next 10 yrs
Capacity purchases	By individual planned firm purchases	$/kwh	¢/kwh	Each yr for the next 10 yrs
Capacity retirements	By individual generating units			Each yr for the next 10 yrs

NOTES: From Randi Lornell, "A PURPA Primer," *Solar Law Reporter*, May/June 1981, p. 45.
The data in this table is required by PURPA Section 210 of all utilities with sales of at least 500 million kwh annually. Utilities with sales greater than 1 billion kwh annually must have reported these data by 1 November 1980. Utilities with annual sales of between 500 million kwh and 1 billion kwh must have reported by 30 June 1982.

hydro at the Energy Law Institute in Concord, New Hampshire. He told Congress in hearings held a year after PURPA's legislated implementation date, "It is important to note that FERC, the state regulatory authorities, and unregulated utilities were, for the most part, unfamiliar with the process of implementation. In the first place, state regulatory authorities were setting wholesale rates and determining interconnection standards in an area where their regulatory authority had seldom, if ever, been exercised. Secondly, state regulatory authorities and unregulated utilities were setting rates for sales of power by qualifying facilities based on the marginal or, as FERC called it, the 'avoided' costs of the purchasing utility (something they also had never done before)."[1]

The FERC itself warned the nation that the PURPA implementation process would involve a giant learning experience. "The Commission continues to believe . . . that this rulemaking represents an effort to evolve concepts in a newly developing area within certain statutory constraints. The Commission recognizes that the translation of the principle of avoided capacity costs from theory into practice is an extremely difficult exercise, and one which, by definition, is based on estimation and forecasting of future occurrence. Accordingly, the Commission supports the recommendation made in the Staff Discussion Paper that it should leave to the state and nonregulated utilities 'flexibility' for experimentation and accommodation of special circumstances with regard to implementation of rates for purchases."[2]

In 1981 the Utah Public Service Commission agreed that the development of rules governing this new relationship between utility and customer must be an evolutionary one. "The concept of avoided cost-based rates is new to this commission," it noted, "and does present difficulties that are yet to be satisfactorily resolved. The Commission therefore intends this order to be a starting point for a process of change and refinement that will at some future point yield proper relationships between utilities and qualifying facilities."[3]

The utility industry's engineers also conceded their lack of experience and definitive knowledge. *Power* magazine, a trade journal for electrical engineers, counseled its readers that, when trying to protect the grid system and the new power producer, "right now, no one, including your utility, knows how much protection [from low-quality energy inputs] is prudent, and how much is overkill, in your specific interconnection situation."[4]

This lack of certainty has tranformed the nation into a giant laboratory. Thousands of experiments are going on simultaneously. Different states and even different utilities within states have widely varying buyback rates and interconnection standards. Sometimes the small power producer can shop around even within small geographic areas. One hydro developer in Oregon found that his property was bounded on one side by a rural electric cooperative (REC) and on another by an investor-owned utility (IOU). Each utility offered different contractual terms. The REC offered lower rates but a better long-term contract than the IOU. He was still deciding which utility to go with when this book went to press.

Qualifying for PURPA

PURPA recognizes two types of qualifying facilities: small power producers fueled by renewable sources of energy and cogenerators. No small power production facility can be larger than 80,000 kilowatts (kw). That includes the capacity of any other facilities owned by the same person and using the same energy source at the same site. (However, producers with between 30,000-kw and 80,000-kw capacity are subject to certain provisions of the Federal Power Act and must have their rates approved by the FERC.) At least 75 percent of the fuel a small power producer consumes in any calendar year must be from biomass; waste; renewable resources like sun, wind or water; or geo-thermal steam or any other combination thereof.

Cogenerators can be fueled by natural gas or oil as well as by such renewable resources as wood and alcohol produced from biomass. There is no size limitation on cogenerators. However, they must meet certain operating standards. Cogenerators are of two types: topping cycle and bottoming cycle. A *topping-cycle* cogeneration facility first produces electricity and then uses the waste heat from this process to provide useful thermal energy. A *bottoming-cycle* facility reverses the process. It first produces useful thermal energy and then uses the waste heat to produce electricity. Almost all the cogeneration equipment discussed in this book uses the topping cycle, because a bottoming-cycle facility that is under 200 kw is at this time uneconomical.

The useful thermal energy produced from a topping-cycle co-generator must be no less than 5 percent of the total energy output in any calendar year. This is meant to prevent a person from installing a generator and only producing electricity, thereby not capturing the efficiencies of cogeneration.

If the topping-cycle facility uses any oil or natural gas and was installed on or after 13 March 1980, it must meet an additional effi-ciency standard. Once again the purpose is to encourage high effi-ciency. In this case, if the useful thermal output of the cogenerator is less than 15 percent of the total energy output, then the electric output plus one-half the useful heat recovered must be at least 45 percent of the total oil and gas consumed. This means if one recovers a relatively small proportion of the waste heat, the QF must convert a higher portion of the energy consumed into electricity than otherwise would be necessary. If only 10 percent of the energy consumed is recovered as heat, then the facility must convert at least 40 percent $(45 - 10 \div 2)$ of the oil or gas consumed into electricity. If the QF recovers at least 15 percent of the total energy consumed as useful heat, then the combination of electric output plus one-half of the useful thermal energy need only be 42.5 percent of the oil and gas consumed in the process.

A cogeneration facility can have a dual fuel capacity and still qualify for PURPA benefits. Such a facility can automatically switch from gas to oil or another fuel source.

Other provisions require that not more than 50 percent of either the small power producer or the cogenerator may be owned by an electric utility, electric utility holding company or a subsidiary thereof. Public utility holding companies that do not own electric utilities may own QFs, as may electric utility holding companies and subsidiaries that are not'primarily involved in generating and selling electricity. If the primary fuel source of the QF is geothermal, 100 percent of the facility may be utility-owned.

Applying for qualifying status is essentially a simple process. Any facility that meets the criteria previously discussed can "self-qualify" (except for dual fuel cogenerators). Or, alternatively, the facility can request a formal certification by the FERC. The latter takes time but has certain legal and regulatory advantages. The utility knows you have contacted the FERC. It may presume you have a slightly higher level of sophistication than the self-certifier and may bargain with you accordingly. In the case of hydro developers, one is given certain powers of eminent domain to make small-scale hydroelectric systems viable, and a license gives you exclusive rights to develop a site for a certain period of time.

The FERC requires the following information:

1. The name and address of the applicant and location of the facility
2. A brief description of the facility, including a statement indicating whether such facility is a small power production facility or a cogeneration facility
3. The primary energy source used or to be used by the facility
4. The power production capacity of the facility
5. The percentage of ownership by any electric utility or by any public utility holding company
6. The location of the facility in relation to any other small power production facilities within one mile of the facility owned by the applicant, which uses the same energy source

Upon receiving an application, the FERC will issue an order granting or denying the application or scheduling consideration of the application. An order denying certification must identify which requirements were not met. If the FERC does not issue an order within 90 days of the filing of a complete application, qualifying status is deemed to have been granted by the FERC.

The FERC may revoke the qualifying status of a facility that fails to comply with the statement contained in its application. So, a qualifying facility planning to undertake substantial alteration or modification may choose to apply to the FERC for a determination that the proposed changes will not result in a revocation of its qualifying status.

As of December 1982 the FERC charged no significant fees for certifying projects. However, proposed fee schedules for various services were issued in late 1982 for comment. The FERC proposed to charge any QF $2,600 to have its project certified. Moreover, the FERC proposed to charge additionally if the QF appealed to it to gain approval to interconnect with and sell power to a utility. The charges range from a minimum of $6,200 to a maximum of $57,400 if the interconnection is contested by the utility and hearings are needed. Critics argue that such fees undermine the intent of PURPA to encourage cogeneration and renewable-based electric generation. Even if such fees were justified on the basis of the FERC's internal costs, critics maintain that they should not be levied on small producers. Those with projects of less than 100-kw capacities would delay or cancel investments in the face of such stiff charges. In early 1983 the FERC delayed its final decision on implementing these new regulations until late 1983.

Utility Avoided Cost Data

The heart of PURPA is the requirement that utilities pay QFs a price based on their own avoided costs. Much of the data upon which that price is set comes from the utilities themselves. The FERC requires utilities with retail sales of more than 500 million kilowatt-hours (kwh) to provide data at least every two years from which avoided costs can be derived. State utility commissions or publicly-owned utilities can ask for more frequent data submissions. Each regulated utility must give these data to the state commission. Both regulated and unregulated utilities must maintain them for public inspection.

Utilities must give the following data to state utility commissions:

1. The system's estimated avoided cost of energy for various levels of purchases from QFs. These levels must be given in blocks of not more than 10 percent of the system peak, or 100 megawatts (Mw), whichever is less. For example, a utility that has 50,000 kw of system peak might serve 5,000 people. It must determine its avoided costs if QFs displace 5,000 kw of capacity. The avoided cost must be stated in cents per kilowatt-hour during daily and seasonal peak and off-peak periods, by year for the current calendar year and for each of the next five years.

2. The utility's plan for the addition of capacity, the amount and type for purchases of firm energy and capacity it

has with other utilities and for capacity requirements for each year during the succeeding ten years.

    3.   Estimated capacity costs, at completion of the planned capacity additions and planned capacity firm purchases (purchases of capacity a seller must deliver at the buyer's request) in dollars per kilowatt and the associated energy costs of each unit, in cents per kilowatt-hour. These costs are to be expressed for individual generating units and purchases.

Appendix 5 presents cost-of-service data for several utilities. Acquiring this type of data is the first step in estimating buyback rates. Notice the difference not only in the estimated avoided costs but in the manner in which each utility presents the data. Seattle City Light indicates that some of its figures are in current dollars, but certain energy cost escalation rates are assumed. In other cases, the costs are in 1981 or 1972 dollars. Seattle City Light could not provide capacity costs, due to confusion over the FERC regulations. Houston Lighting and Power provides costs in current dollars. It gives no escalation rate. It gives a breakdown of avoided costs for 100 Mw, 500 Mw and 1,000 Mw.

The regulations do not specify a methodology for determining avoided costs. That is left up to the state regulatory agencies and the unregulated utilities (via their city councils) or rural electric cooperative boards. They do, however, provide guidelines, which are discussed below.

Utilities with retail sales of less than 500 million kwh, other than an all-requirements utility, must provide data comparable to those described in Appendix 5 upon request, so that QFs can estimate the utility's avoided cost. It is important to remember that, in most cases, the public utility commission does not establish actual rates. Rather, it establishes the methodology with which to calculate the actual rates. However, the utilities must develop a standard contract for QFs with capacities under 100 kw, and most commissions monitor this closely and will intervene to establish rates directly if the utility proves recalcitrant or is viewed as promoting inappropriate tariffs. An *all-requirements utility* is one that purchases all of its electricity from a bulk supplier. It may provide the data of its supplying utility and the rates at which it purchases energy and capacity.

Standard and Negotiated Contracts

To minimize the expense of negotiations, PURPA requires that utilities offer a standard contract to QFs with capacities under 100 kw.

The standard contract is a useful device to balance the inherently unequal bargaining power of the utility and the small power producer. As Paul Gipe, an adviser to small wind machine owners in Pennsylvania says, "Small power producers come to the utility hat in hand."

States can also require standard contracts for generators of more than 100 kw, although PURPA does not require this. New Jersey, for example, requires a standard contract for all QFs with a capacity of less than 1 Mw (1,000 kw). Montana requires a standard contract for all QFs regardless of size. The standard contract should contain the minimum provisions the regulatory commission believes necessary to attract the maximum number of QFs. Standard contract prices must reflect full avoided costs.

The owner of a power plant with less than 100-kw capacity (or the state minimum) can choose the standard contract or can negotiate individually for a better deal. QFs with capacities greater than the state minimum required for a standard contract must negotiate individually. PURPA still provides the guidelines for individual negotiations, but it is not the final word. As one handbook for small hydro producers advises, "Thus, while PURPA has become a vital 'silent partner' of private power producers, it is largely still up to the individual hydro developer to secure a satisfactory agreement through a process of negotiation with a purchasing utility. Following a thorough understanding of his own costs and requirements, as well as those of the prospective purchaser, the producer will, at an early point in his negotiating strategy, be confronted with the need to establish very precise terms and conditions with respect to the proposed transaction. For it is within the framework of these provisions that the most critical bargaining will take place."[5] One point to be stressed here is that it is legal under PURPA for a QF to negotiate any deal or set of contract terms with a utility. PURPA laws are not a limiting framework, and the QF is free to try to obtain the best contract terms possible.

QFs unhappy with specific contract provisions offered by the utility or who believe the utility is delaying the negotiation process have the right under PURPA to appeal to the regulatory commission or the FERC. Although no states have yet developed clear appeal procedures, some QFs have already used their regulatory commissions to good effect. Sometimes all it takes is a phone call to the commission staff, who then calls the utility to investigate the matter informally.

Introducing the Actors

Small power producers and cogenerators with capacities of less than 200 kw are the primary audience for this book. This size range embraces the 2-kw rooftop photovoltaic array as well as the medium-size motel cogeneration system. Potential power producers usually fall into one of three categories.

First is the producer who wants only to sell electricity. The owner of a 50-kw hydropower facility or wind farm has no purpose other than to sell electricity to the grid system. The project's cost-effectiveness depends on the buyback price and the cost of the interconnection equipment. If the buyback price is too low, the project will not be developed. This facility needs no back-up power. Many of these projects will be financed by outside investors who need long-term contracts and some certainty of future purchase prices to encourage their investment.

The second type of producer uses some of the electricity on-site and sells only the surpluses. For this person, the price of utility back-up power will become a very important factor. The size of the power plant may vary depending on the price the utility pays for the delivered power. This is especially true for cogeneration systems. The cogenerator can choose, if the buyback price is too low, to size the system to meet only the internal heat load, with the electricity viewed as an additional benefit. If the buyback rate were higher than the retail cost of electricity, the size of the system might be increased to take advantage of the additional revenue from sales to the grid. According to Richard Nelson of Cogenic Energy Systems, a major supplier of small-scale cogeneration systems (see Appendix 6), none of the more than 100 systems installed by his company were sized to deliver electricity back to the grid.

The third type of producer generates only for internal consumption. A 2-kw residential solar cell array could represent such a case. This producer may have no plans to sell electricity to the utility. The cost of two meters and certain interconnection equipment may outweigh the revenue possible from sales. This producer may not be interested in buyback rates but in the provisions of PURPA that prohibit high back-up or other service charges.

Major Negotiation Issues

Many QFs lack knowledge about their rights under PURPA. Sunday supplements carry heartwarming stories about the family that now owns a small hydroelectric plant, but they rarely give any detailed information about how the system fits into the overall electric system. While interviewing people for this book, we discovered many examples of ignorance being harmful to the independent producer. One owner of a photovoltaic array in California didn't know that she qualified for capacity payments under the standard contract even though sunlight is an intermittent source of fuel. A hydroelectric developer in Utah wasn't aware that even though his state's major utility had an excess capacity fired by coal, this didn't rule out a high avoided cost and therefore a high buyback rate.

In the bargaining process, the potential QF must be ready to deal with a wide variety of issues. Listed are the major ones in the form of questions. Each will be addressed in the course of this chapter.

What was the purchase price based on? How often could it be changed? Is net billing available? Are long-term contracts available? Do such contracts include provision for rising purchase prices if energy costs increase? Is there any protection against declines in future avoided costs? Are levelized payment schedules offered for long-term contracts?

Does the utility offer time-differentiated rates? Is the difference between peak and off-peak buyback prices sufficient to compensate for the additional cost of a time-of-day meter?

Are capacity credits available? How much are they worth? How are they estimated? How often can they be changed? What is the minimum length of contract the QF has to sign in order to qualify for capacity credits? Are capacity credits available on an "as available" basis or does the QF have to meet certain performance standards including availability during peak periods to earn capacity credits? Is a penalty attached to a QF failure to meet these standards?

Under what conditions can the utility cease purchasing power from a QF? In such instances what notice does the utility have to give the QF? What justification does the utility have to present to the regulatory commission?

Are customer service charges imposed on the QF in addition to those imposed on other customers in the same class who do not generate power? Are back-up or standby rates different for QFs than for other customers? Have utilities justified these differences on the basis of differences in their cost of service?

How often can the contract be renegotiated by the QF or the utility? When must notice of termination be given by either party?

Does the utility include QFs as part of its resource plan? What is the total capacity of QFs it plans to have on-line within the planning horizon?

How does one appeal to the public service commission or, in the case of the unregulated utilities, to the utility board or the city council? How long does an appeal take?

Does the utility require the QF to indemnify it for possible damages? Does indemnity work both ways? Must the QF take out an insurance policy to cover possible injuries sustained to the system or to other customers? Is a specific amount of insurance coverage indicated?

Are there provisions in the standard contract offered to applicants under 100 kw that are not in sample negotiated contracts?

What is the responsibility of the utility to wheel QF-generated electricity to a neighboring utility that might pay more for the electricity? What is the charge for such wheeling?

If this list of questions appears mind boggling, don't worry. In the beginning you will have a new jargon to learn. You will have to learn to think systematically. But eventually the answers to individual questions will form a picture. Most relate to others. You are discovering the relationship of the individual producer to the electric grid system. Remember, while you are going through the process, so are dozens, even hundreds of others. Several states have their own associations of independent power producers. Link up with them. Form your own. Share information. Piggyback on their experiences. You can do it.

Those who master the process may get their just rewards. In several states individuals who negotiated knowledgeably more than doubled the price they received for their electricity.

Although most of this book is devoted to providing technical information, one point cannot be stressed enough. The process of PURPA proceedings is a political one. Clearly there is a rational, technical, mathematical basis for the final decision. But the final decision can vary dramatically depending on the political clout of the small power producers. Examples of this abound. Utah Power and Light (UP&L) serves areas of Idaho as well as Utah. Most of the data upon which it based its avoided costs in both states was identical. Yet Idaho concluded that UP&L should pay 4.6¢ per kilowatt-hour and Utah concluded it should pay about 2.4¢. Needless to add, independent power producers intervened aggressively in Idaho's hearing process. Utah had little participation during its initial process.

The large issue of interconnection requirements has not been raised here. That issue is so complex that the entire next chapter is devoted to it.

Purchase Prices and Avoided Costs: A Brief History

For many small-scale power producers, the key to PURPA is the rate the utility must pay for independently produced electricity. PURPA requires utilities to buy energy from QFs at rates based on the utilities' incremental costs. Thus the purchase prices (or buyback rates) should be based on the cost of the next unit of energy or capacity that the utility would have to bring on-line. As discussed in the last chapter, the utility *sells* electricity at its *average* rather than its *marginal* cost. If it must buy electricity at its marginal or future costs, it may pay a QF a higher price than it charges. This is especially true if the QF can receive a capacity credit. This favorable situation depends greatly on aggressive public service commission involvement in rate structuring. In early 1983 this was the case in most parts of Oregon, Idaho, Montana, New Hampshire and Vermont.

To most QFs, no issue is more important than the buyback rate. Today these rates vary by more than 5:1 around the country. A 4-kw wind machine that generates 9,000 kwh a year in 12-miles-per-hour (MPH)-average wind would receive 1.2¢ per kilowatt-hour in mid-1982 in Nebraska and almost 8¢ per kilowatt-hour in Vermont. That represents a range in first-year revenue of $100 to $700. In the first instance, assuming a 10 percent annual increase in the buyback rate, the machine will still not repay its initial investment over its life. In the second case, it pays itself off in about seven years. PURPA itself does not establish a specific buyback rate. Congress said only that purchase prices (1) shall be just and reasonable to the electric consumers of the electric utility and in the public interest, and (2) shall not discriminate against qualifying cogenerators or qualifying small power producers. In addition, "No such rule prescribed under subsection (a) shall provide for a rate which exceeds the incremental cost to the electric utility of alternative electric energy." Congress simply defined "incremental cost of alternative electric energy" as "the cost to the electric utility of the electric energy which, but for the purchase from such cogenerator or small power producer, such utility would generate or purchase from another source."[6]

The Federal Energy Regulatory Commission translated this relatively broad mandate into specific regulations. Its regulations put meat on the bare legislative bones. The FERC redefined "incremental costs" as "avoided costs" and ordered utilities to pay 100 percent of their avoided costs to QFs. To the FERC, these represented those costs that "but for the purchase from a qualifying facility, the electric utility would generate or construct itself or purchase from another source."[7]

To the electric utilities, the FERC regulations constituted a declaration of war. The utility-QF relationship has been from the beginning an adversarial one. Most utilities view QFs as potential competitors. Utilities believe QFs even under ideal conditions can generate only insignificant quantities of power in the foreseeable future. Therefore they do not believe that QFs can actually displace a multi-megawatt power plant.

If this interpretation were accepted by utility commissions, it would become a self-fulfilling prophecy. The utility concludes that QFs can never produce sufficient power to displace a central power plant. Therefore it offers no capacity credit. As a result, few QFs enter the market because the economics of doing so are marginal. The FERC turned a deaf ear to the utilities' forecasts and ordered them to pay QFs a capacity credit in return for a contract for firm power (that is, power generated a certain fraction of the year) as if the individual QF were displacing a piece of a future generating plant because the aggregate class of a type of QF (e.g., all the wind-electric systems, all the hydro-electric systems and so forth) does itself provide some level of firm power.

Utilities also vigorously disputed the FERC's requirement that they pay QFs their full avoided costs. They argued that, if QFs could generate power at very low costs, the utilities should be given the ability to bargain flexibly so that a portion of the savings could go to their ratepayers. They argued further that by denying them the right to share in the immediate benefits of small power production, the FERC undermined any incentive they might have to encourage dispersed generation actively. In Congress, before state regulatory commissions and in federal courts, utilities consistently presented the case for full avoided costs as a ceiling rather than a floor.

To utilities, having to pay 100 percent of their avoided costs meant giving a windfall profit to QFs. They argued that the first QFs would install dispersed generation plants on the most attractive sites, harnessing the fastest flowing rivers equipped with existing dams, or the windiest terrains, or industrial cogeneration systems in businesses that could use a large portion of the waste heat. Utilities expressed chagrin that investors in dispersed generators could earn a return of 30 to 50 percent or more while they were restricted to a regulated return of 11 to 17 percent.

Utilities argued that high purchase prices were unnecessary since many dispersed generation facilities would be profitable even if they received a lower price. Why pay full avoided costs when the same amount of capacity could be brought on-line at a lesser cost to the ratepayers?

Finally, utilities argued that they should not be denied PURPA benefits. The FERC restricted utility ownership to no more than 50 percent of individual QFs. This restriction once again reduced the utilities' incentive to encourage small power production and cogeneration. If Congress really wanted to encourage electrical generation from such technologies it should allow utilities, who after all know the electric generation business better than anyone, the same incentives it does private unregulated industries.

The FERC denied each of these arguments. It admitted that paying full avoided costs might give those QFs that acquired the most attractive facilities a very high return on their investment. But the FERC argued that this rapid penetration was exactly what Congress intended, and that these pioneers were taking the risk of dealing with relatively new technologies and new regulatory and legal requirements, not to mention recalcitrant utilities.

The FERC further argued that to base the purchase price on the return the QF would be getting would require a detailed examination of the QF's production costs. Such an examination smacked of the kind of conventional utility regulatory investigations that were the sort of bureaucratic deterrent Congress wanted to avoid.

Finally the FERC argued that paying full avoided costs encourages not only those at the tip of the iceberg to invest in small power production and cogeneration but all of the iceberg that is below the avoided-cost waterline. Assume that a utility's next power plant would generate electricity at a cost of 9¢ per kilowatt hour. QFs that can generate electricity at a production cost of 3¢ per kilowatt hour might still invest even if they were paid a fraction of the full avoided cost. But investments in QFs that generated electricity at 6¢, 7¢ or 8¢ per kilowatt hour would not be profitable unless the investor received the full avoided cost. Yet these facilities would nonetheless have provided cheaper electricity than the utility's. It was to gain the active participation of this "second tier" of QFs that full avoided-cost payments were proposed.

Indeed, several state agencies argued that full avoided costs actually represented a conservative price. The California Public Utility Commission commented, "If anything, this Commission believes that the avoided cost signal to QFs is a conservative one, as it does not include the tangible but hard-to-quantify 'social costs' that are associated with new utility supplies and which are avoided through the purchase of QF power. These 'social costs' include the risks associated with imported energy supplies and environmental degradation related to conventional generation."[8] In short, at least one regulatory body saw in QFs a value beyond just the utility's avoided cost. Avoiding the pollution of a new coal plant or the waste problems and potential danger of a new nuclear facility were also of value.

Ironically, at the moment the commission issued this statement, one of its regulated utilities, Southern California Edison, was requesting permission to charge its ratepayers $3.15 million over six years for its share of the $760 million remaining cleanup cost of Three Mile Island. No more perfect example of the social costs of conventional electric generation could have been given.

In fact, the FERC does not prevent state regulatory commissions or unregulated utilities from establishing purchase prices that are higher than avoided costs. It prohibits only the setting of purchase prices lower than avoided costs. Both Montana's and New Hampshire's regulatory commissions have taken public note that their purchase prices may sometimes be higher than full avoided costs to encourage greater participation of QFs.

The FERC also denied the utilities' request that they be permitted to own QFs. It noted that nothing in PURPA prevents a utility from owning wind turbines or cogeneration facilities. PURPA only denies such facilities eligibility for the unique PURPA benefits. Utilities *should* be building wind turbines and hydro plants where they are cost-effective. Moreover, before 1981 and PURPA's implementation, utilities had shown *no* desire to build cogeneration plants or to install small hydro or wind facilities. The FERC believed that this lack of initiative demonstrated their lack of commitment. Finally, the FERC hoped and expected that the field of cogeneration and small power production would be highly competitive. Having operated in a highly regulated, monopolistic environment, the utilities probably had little expertise or inclination to operate in such a field. If a subsidiary of the utility owned a QF, there arose the additional possibility of collusion or at least favoritism.

To understand properly the PURPA negotiation process and the intense controversy surrounding the FERC's regulations, one must realize the extent to which PURPA reverses the conventional utility role. Traditionally the utility has exercised a monopoly over electric power generation and sales. Any monopoly tries to maximize the price it receives and the profit it makes. That is why regulatory commissions were established to represent the public interest and the ratepayer. PURPA eliminates the monopolistic role, but it replaces it with a *monopsonistic* one. PURPA does not allow QFs to sell at retail, preserving the monopoly by utilities over transmission and distribution and thus creating monopsony. A monopsony occurs when there is only one buyer. Before PURPA the utility was the only seller. Under PURPA it is the sole buyer but not the only seller. Under monopsony the buyer's objective is to pay the lowest price.

Thus one can expect an adversarial role even between the most progressive utility and the potential QF. Even when a utility wants to encourage QFs, it will still try to buy electricity on the best terms for itself. When a utility is anything less than sanguine about the rise of independent power producers, its negotiation position can be downright discouraging. A full discussion of negotiating strategies and examples is the focus of chapter 5. However, any QF should have a basic understanding of the economics of utility electric generation. Knowledge is your best ammunition. What follows is a discussion of the basic concept of avoided cost and the various ways that this cost is determined.

Calculating Avoided Costs

Dispersed electric generation allows the utility to avoid several kinds of expenditures. One is the expenditure for fuel to fire the plant that would ordinarily carry the top of the utility's load at the moment the QF is producing power. Another is a modest savings in the cost of the plant due to reductions in operation and maintenance expenses because of a decreased annual use. These types of savings are called *short-term avoided energy costs*. The second type of potential savings are *long-run energy costs*. These savings can also include a capital credit if, as is the case for most utilities, the management policy has been to substitute capital for fuel by building nuclear (low fuel cost, high capital cost) rather than oil-fired (higher fuel cost, lower capital cost) power plants. The third type of avoided cost is the savings in not having to construct new power plants. These are called *capacity credits*. Another avoided cost is in savings to the transmission and distribution system because the dispersed generator's output will be consumed nearby, thereby lessening the load on the utility's transformers, voltage regulators and lines.

To the FERC, "the costs which an electric utility can avoid by making such purchases generally can be classified as 'energy' costs or 'capacity costs.' Energy costs are the variable costs associated with the production of electric energy (kilowatt-hours). They represent the cost of fuel and some operating and maintenance expenses. *If, by purchasing electric energy from a qualifying facility, a utility can reduce its energy costs or can avoid purchasing energy from another utility, the rate for a purchase from a qualifying facility is to be based on those energy costs which the utility can thereby avoid.* Capacity costs are the costs associated with providing the capability to deliver energy; they consist primarily of the capital cost of facilities. If a qualifying facility offers energy of sufficient reliability and with sufficient legally enforceable guarantees of deliverability to permit the purchasing electric utility to avoid the need to construct a generating unit, to enable it to build a smaller, less expensive plant, or to purchase less-firm power from another utility, then the rates for such a purchase will be based on the net avoided capacity and energy costs."[9] (Emphasis added by author.)

Short-Run Avoided Energy Costs

Utilities operate several kinds of power plants that vary in construction cost and operating expense. Coal or nuclear-fired baseload plants might have an operating cost of between 1¢ and 2¢ per kilowatt-hour. Coal is cheap. In 1982 a ton cost $40.00, or about $1.35 per million Btu. Assuming a heat rate of 10,000 Btu per kilowatt-hour, the cost of electricity would be 1.3¢.

Oil, on the other hand, is expensive. Oil-fired peaking plants can cost 7¢ to 15¢ per kilowatt-hour to operate. Thirty-three dollars a barrel translates into $5.69 per million Btu. Assuming a heat rate of 12,250 Btu per kilowatt-hour, the kilowatt-hour cost is about 7.2¢ (Natural gas falls between coal and oil. In 1982 it cost 33¢ per cubic foot or $3.30 per million Btu. Deregulation should produce parity between oil and natural gas by the mid-1980s.)

Remember, *PURPA requires that the QF be paid a price equal to that of the most expensive plant operating when the QF generates electricity.* The QF is displacing the top of the load, the peak. New Hampshire's Public Service Commission, for example, concluded that even though its utilities rely heavily on nuclear power, the top of the load (demand) should almost always be supplied by oil-fired power plants. Thus the QFs are paid a high price based on the costs of operating oil-fired plants. (If the Commission had concluded that nuclear plants were able to meet 100 percent of the demand during a significant period of time, the utility would have had to pay QFs only about 1.5¢ per kilowatt-hour.)

Many states rely on sophisticated computer models to determine their avoided costs. Others choose a proxy, or stand-in power plant, to provide the data. New Hampshire used the Newington plant, the most efficient baseload, oil-fired plant operated by the Public Service Company of New Hampshire. Since it was oil-fired the fuel was expensive, but it was an efficient baseload plant, so it burned less fuel to produce the same amount of electricity than would a peaking plant.

The commission determined a base avoided fuel cost of 61.18 mills per kilowatt-hour. This assumed a heat rate of 10,250 Btu per kilowatt-hour and an oil price of $35 per barrel. The New Hampshire Commission used that as a starting point for estimating avoided costs. In response to a utility petition, the commission stated, "The position that avoided costs should be based solely on average fuel costs is rejected." The commission then raised the base fuel cost to account for the time the efficient Newington plant would not be available. When the Newington plant was not available (e.g., during periods of forced outage) or when the system load exceeded the total capacity of Newington to carry the top of the load, more expensive peaking plants would be used. These plants also burn oil but more inefficiently. Their higher heat rates mean that the cost of producing a kilowatt-hour rises. The commission increased the avoided costs by 17 percent to reflect "the unscheduled outage rate at Newington and the weighted cost of all units more expensive than Newington."[10]

The New Hampshire Commission was aggressive in searching out other savings gained from the inputs of QFs. For example, it increased the buyback rate 0.2¢ per kilowatt-hour to account for the displaced working capital that utilities would otherwise have had to use to pay for fuel inventory. Another 0.21¢ per kilowatt-hour was tacked on for reduced operation and maintenance costs. The lessened use of the Newington station meant reduced deterioration and therefore a longer useful life.

These additions increased the base fuel cost of 6.18¢ to a purchase price of 7.63¢ per kilowatt-hour.

New Hampshire's buyback rate does not vary throughout the year. Most commissions have rates that vary by time of day and season. California is one state that breaks down the day into three rating periods. Table 3–3 shows the incremental heat rate and energy purchase

TABLE 3–3

**Calculation of PG&E's Energy Purchase Prices
for November and December 1981 and January 1982**

TIME PERIOD	AVERAGE INCREMENTAL HEAT RATE (BTU/KWH)	UNIT COST OF ENERGY CONTENT ($/10^6 BTU)*	ENERGY PURCHASE PRICE (¢/KWH)†
On-peak	11,850	6.5193	7.725
Partial peak	11,200	6.5193	7.302
Off-peak	10,000	6.5193	6.519
Annual average	10,860	6.5193	7.080

SOURCE: PG&E, 1981.
* Cost of oil received ÷ energy content of oil received.
† Average incremental heat rate × unit cost of energy content.

price for Pacific Gas and Electric (PG&E) for each rating period and its annual heat rate and energy cost. To understand how PG&E derived the 7.725¢ per kilowatt-hour price it was willing to pay QFs during the summer peak hours, take 11,850 Btu per kilowatt-hour and multiply by the $6.5193 per million Btu cost of oil at the time ($32.86 per barrel). Then divide by a million and you get 7.725¢.

Table 3–3 displays the computation of energy purchase prices to be offered to qualifying facilities in November and December 1981 and January 1982. This table is based upon the unit cost of energy content value of $6.5193 per million Btu and the average incremental heat rates for various pricing periods.

California's system is complicated inasmuch as it has many different rating periods and is based on PROMOD, a highly sophisticated computer model that the utility industry uses for regional power pool dispatching. Some controversy has been generated about exactly what type of plant is carrying the top of the load curve at any given time period. Since the incremental heat rate varies by some 18 percent, from 10,000 Btu to 12,340 Btu per kilowatt-hour, the controversy is not trivial. A 100-kw cogenerator operating at a capacity factor of 80 percent would generate about 500,000 kwh a year. The 1.2¢ per kilowatt-hour difference in heat rates equals a difference of $6,000 in annual income to the QF. However, the difference will only be that great if the 1.2¢ variation occurs during the entire year. That is unlikely. It would probably occur only when peaking plants are operating during off-peak or mid-peak times. Thus only a fraction of the $6,000 is at stake.

Often utilities offer nonfirm energy purchase prices and time-differentiated energy prices. The nonfirm price is usually an average of peak and off-peak. In 1982 the Oregon-based Pacific Power and Light Company (PP&L) offered a price of 2.28¢ per kilowatt-hour, or a time-differentiated price of 2.76¢ on-peak, 6 A.M. to 10 P.M. Monday through Friday, and 1.84¢ at all other times. The individual QF in this case would benefit from the time-differentiated prices only if it can generate a great deal more power on-peak than off-peak. But the QF would have to install a meter for each time period. Thus in California a three-register meter must be installed. Each additional register costs $50. If the amount of power the QF can generate during peak periods is greater than the cost of installing extra meters, only then is the time-differentiated rate beneficial to the QF.

In some cases, the average price will not be midway between the off-peak and peak rate. The Omaha Public Utility District offered 1.6¢ at peak in 1982 and 1¢ at off-peak. The nonfirm rate was 1.1¢ per kilowatt-hour. Since a time-of-day meter costs $100 to $300, the QF must be able to deliver 2,000 kwh to 6,000 kwh annually on-peak just to balance the cost of the meter. (Typically there are 2,000 peak hours a year.) A 5-kw wind machine operating at 30 percent capacity would generate 13,140 kwh per year. If 20 to 50 percent of this output were generated during the utility's peak periods, the time-of-day meter would be a good investment.

Calculating short-term energy costs is relatively easy. The major arguments relate to which plant is carrying the top of the load curve and what associated costs (e.g., displaced working capital, line losses and so forth) the QF displaces.

Long-Run Avoided Energy Costs

The concept of long-run energy costs is more difficult to understand. Many utilities are "backing out" of oil-fired power plants by substituting coal and nuclear plants. That is, they are substituting baseload plants for peak load plants. Thus when the QF displaces the top (peak) of the load, it may be displacing inexpensive coal or uranium fuel rather than expensive oil or natural gas. The baseload plants are expensive compared to peaking plants, but coal and uranium are inexpensive fuels. Utilities argue that avoided energy cost is the cost of operating these plants and that they should be required to pay QFs only this amount. Since coal- or nuclear-powered baseload plants have very low operating costs, this interpretation of avoided costs would lead to very low buyback rates. Where this interpretation has been accepted by public utility commissions, as in Nebraska and Iowa, this has been the case.

If the utility argument is accepted, the QF is placed in a difficult situation. The utility that substitutes base load for peaking plants will tend to have an excess capacity, thereby reducing the availability of capacity credits to the QF. If the avoided energy costs are based on the price of coal or uranium, the QF will receive a low price for its electricity. As was mentioned before, some regulatory commissions designed an as available energy tariff under which the QF receives a higher kilowatt-hour rate for long-term firm power contracts. These as available rates reflect a certain amount of avoided capacity. If the QF wants to receive a higher capacity credit, then it must meet the higher performance standards set by the utility for these credits.

But other commissions located in states with coal-fired plants do not interpret the energy costs in so restricted a manner. They assume that by building more expensive coal and nuclear plants the utility is substituting capital for fuel. The plant is not being built to meet future growth in demand but rather to substitute for oil- or natural gas-fired plants. Thus the portion of the plant's cost that exceeds the cost of a combustion turbine ($300 per kilowatt in 1980) can be interpreted as an energy-related rather than a capacity-related investment.

If this is done, the long-run avoided energy cost for a typical $2,000-per-kilowatt coal plant is about 5.2¢ per kilowatt-hour. Assuming $1,700 of the plant's cost per kilowatt is related to energy and it operates at an 85 percent capacity factor and the fixed capital charge is 17 percent, the capital-related, energy-cost component alone would be 3.9¢ per kilowatt. A typical nuclear plant would have an even higher energy-related payment.

Montana's Public Service Commission has adopted this methodology. In return for gaining capital-related energy payments, the QF must sign a long-term contract but need not meet any reliability or operating standards that would be associated with capacity credits. In Montana in 1982, QFs with no contract received a short-term energy rate of 2.1¢ to 2.3¢ per kilowatt-hour, depending to which of the three Montana utilities they sold their electricity. For those QFs willing to sign a minimum four-year contract, both a long-term energy rate and a long-term capacity credit applied. Capacity credits will be discussed shortly.

The estimation of long-term energy rates requires the use of modest arithmetic. The Montana formula for the long-term energy rate (LTE) follows:

$$\text{LTE} = \frac{\left[\left([a\,(c+e)] - [b\,(d+f)]\right)(1+g)\right]}{(8{,}760 \times 0.70)} + \frac{[(h)\,(j)]}{i} + k$$

This formula will be explored in detail because the Montana methodology is a model that could very well be used by the entire country. The formula looks formidable, but once the letters are defined, the equation loses its mystery. The baseload capital costs are represented by *a*, and the baseload annual carrying charge by *c*. This is called the *fixed-charge rate*. When you take out a loan and want to know how much you must pay each year to repay the loan principal plus interest, you multiply the loan by the fixed capital charge rate. That rate varies depending on the interest and the term of the loan. The baseload fixed operation and maintenance cost are represented by *e*. Thus the first part of the formula calculates the annual costs of repaying the investment in a baseload plant plus the maintenance and fixed operating costs of that plant. The baseload plant will be expensive. In Montana the average cost is over $1,500 per kilowatt.

The next part of the formula deals with the annual costs of a combustion turbine. The capital cost of such a plant is *b*, and *d* is the fixed-charge rate. The rate is higher for this type of plant because the plant is estimated to last only 25 years versus 35 years for a baseload plant. Since the investment must be repaid over a shorter time, the fixed-charge rate is higher (assuming identical interest rates as Montana did). The fixed operation and maintenance costs for the combustion turbine are represented by *f*. The capital cost of the combustion turbine is low. In Montana it is slightly more than $325 per kilowatt.

The annual costs of the combustion turbine are then subtracted from the annual costs of the baseload plant. This then gives the energy-related capital cost. In other words, the assumption is made that a utility is substituting capital for energy by building a coal-fired baseload plant to meet future load. The energy-related capital costs are the additional costs of the baseload plant.

The rest of the formula is easy. The transmission line losses are represented by *g*. Montana assumes an 8.3 percent line loss so that costs determined by the first part of the formula are multiplied by 1.083. This entire figure is then divided by the number of hours in a year, 8,760 multiplied by 0.70, which is a typical capacity factor. This gives the cost in cents per kilowatt-hour for a typical year.

It remains to add the fuel cost and the baseload variable operation and maintenance. The fuel cost is figured out by taking the cost of a pound of coal (*h*) and multiplying it by the heat rate of a baseload plant (*j*) and then dividing this by the number of Btu in a pound of coal (*i*). In the case of Montana, the cost of coal is about $10 a ton or about 0.5¢ per pound. There are about 8,500 Btu in a pound of Montana coal, and the baseload plants have heat rates of slightly less than 11,000 Btu per kilowatt-hour. The fuel cost comes to about three-quarters of a penny. To this is added the baseload variable cost of operation and maintenance (*k*), which comes to about 0.25¢ per kilowatt-hour, and the formula is complete.

The result is a long-term energy rate of from 5.1¢ to 5.3¢ per kilowatt-hour depending on the utility. Thus any QF that signs a contract for a minimum of four years will receive this rate for any kilowatt-hour generated during the contract period.

The Montana case illustrates an important rule. *Long-term contracts allow a QF to receive higher prices for energy. However, in the majority of cases, the price is fixed. In return for certainty about future prices, the QF actually receives a price that decreases in real terms because of inflation.* If the QF can negotiate into the contract a clause that allows the buyback rate to rise with inflation, all the better. This is less a concern for those owning hydroelectric wind power facilities or photovoltaic systems because these are financed by fixed-rate loans. Therefore, the monthly payment is fixed and drops in real terms (because of inflation) as does the buyback price. But for cogenerators who are uncertain about future energy price increases, this can present a genuine problem and make a cogenerator less than enthusiastic about a long-term, fixed-rate contract.

There is always an exception to the rule. In this instance it is the rare utility that experiences no significant variations in daily demand over the course of a year. Economics would justify letting a baseload plant carry the top of the load all the time. There would be no substitution of capital for energy. This might be the case, for example, where there are local industries that operate around the clock or where there is no significant difference between summer and winter demand.

For example, the avoided energy costs would probably be quite low for a utility like Colorado-UTE. It relies entirely on coal-fired generation. The utility has "mine-mouth" coal plants; that is, coal is burned right where it is mined, reducing the cost of the coal by 75 percent (transportation is generally most of the cost of coal). Thus, provided the utility is not planning to build any new coal plants, it can realistically offer QFs only about 0.5 cents per kilowatt-hour for their electricity. (For those interested, the weekly publication *Energy User News* has periodic reports that break down coal costs into mine-mouth costs and transportation.)

Avoided Capacity Costs

No component of the buyback rate has been more controversial than that representing *avoided capacity costs*. This represents capital displaced by deferring or canceling future power plant construction. Utilities often argue that capacity credits are inappropriate for dispersed generators. Their arguments fall into three categories.

First, small power producers like wind machines or photovoltaic arrays produce electricity intermittently. Since they cannot realistically be depended upon to be producing at any specific time of the year, they should not be given credit for displacing capacity.

Second, dispersed generator owners do not hold the long-term public interest as a central consideration. They may abandon the plant 5 years down the line. If that is the case, how can it be said that they are displacing the need for a power plant to be built 10 years in the future and operate for an additional 30 to 40 years beyond that?

Third, dispersed generators are by their nature very small. How can a 10-kw hydroelectric plant be said to displace a 500,000-kw coal plant?

The first and third arguments can be answered together. One 10-kw hydro plant cannot displace a coal plant 50 times its size, but 500 hydro or small-scale power plants can. In the second argument the utility is saying that an individual QF is not motivated by a long-term public interest, which may or may not be true on a case-by-case basis, but the argument has the wrong focus. It is the aggregate class, all QFs taken together, that should be looked at. By virtue of its being a dispersed, growing entity, an aggregate class does in fact have a long-term interest. In its rules, the FERC indicated that rates for purchases shall take into account "the individual and aggregate value of energy and capacity from QFs on the electric utility's system."[11]

In the preamble to its final rules, the FERC notes that "an effective amount of capacity may be provided by dispersed small systems, even in the case where delivery of energy from any particular facility is stochastic."[12] Stochastic means probabilistic. One might not be able to predict how much electricity will come from a specific QF, but one can predict how much electricity will come from a statistical sample of QFs. Thus the wind might be said to be too unreliable to be able to displace any capacity. But many wind turbines located throughout the utility's service area will intercept a minimum amount of wind, and if there are many wind turbines, then the class of wind machines can be said to have a minimum capacity factor. The FERC also notes that "testimony at the Commission's public hearings indicated that effective amounts of firm capacity exist for dispersed wind systems, even though each machine, considered separately, could not provide capacity value. The aggregate capacity value of such facilities must be considered in the calculation of rates for purchases, and the payment distributed to the class providing the capacity."[13]

Measuring precisely the aggregate capacity value of a class of small generators requires that such a class already exists and can be monitored. Obviously, were such a class in existence, PURPA Sections 201 and 210 would be superfluous. Thus, in response to the utility argument that capacity payments be waived until such time as the class exists and proves itself, the FERC orders utilities to estimate the capacity value of technologies under different penetration scenarios. For example, if there were 50,000 10-kw wind machines operating over several hundred square miles, how much of that capacity could be considered firm? Today only an educated guess can be made. One report by Westinghouse concluded that multiple wind turbines should be credited with a 35 percent capacity value. If that figure were accepted and the utility were paying full capacity credits to a facility with a 70 percent capacity factor and the utility commission required partial payments, the wind turbine would qualify for 50 percent of the capacity credit. In five years there should be sufficient operating data to make a much more reliable estimate.

Different technologies have different operating characteristics and, therefore, a different likelihood of receiving capacity payments. Clearly, cogenerators have the best chance. They operate most like a utility power plant. Their forced outage rate is often lower than that of most utility plants. The greater the number, the more reliable they become on a class basis. Indeed, one could argue that, given their greater reliability, cogenerators should be receiving more than a 100 percent capacity credit, which will be discussed shortly.

Outages and breakdowns at one cogeneration plant are not linked in time or associated with a higher incidence of breakdowns at other cogeneration plants. This is not the case with small-scale solar-electric technologies. These technologies have a *correlated output*. All hydro plants operating off the same river and most in the same geographic vicinity will have similar seasonal variations in output due to variations in the river's flow. But these hydro plants will have a minimum firm capacity. If their maximum flow rates coincide with their utilities' maximum demand for power, they can actually be said to displace a higher amount of capacity than if they had a stable year-round output at a lower level.

The output from photovoltaics is clearly intermittent and weather dependent. But here the timing of that output can have substantial benefits. Solar energy is available only during the day; it is less available and less intense during the winter. Yet its availability coincides well with most utilities' peak and intermediate loads. The peak solar radiation in summer coincides with most American utilities' peak air-conditioning loads. Moreover, the output characteristics of solar cells can be shaped by varying the orientation of the arrays. Maximum generation occurs when a solar collector of any kind is at solar noon (between noon and 2 P.M.), but peak generation can be shifted in time at the expense of a reduced maximum overall daily output. The capacity value of solar electric generation can thus be increased to coordinate output to match the system load curve of a utility whose annual peak load occurs at 4 P.M., typical of a summer peaking system.

If the utility is paying a much higher price for peak power than for nonpeak power, or if it pays a good capacity credit for on-peak generation of electricity, it may be worth it for the photovoltaic owner to slope the panels more toward the west or somewhat higher than would be the case if he or she wanted to maximize annual production.

Wind speeds vary across the utility's service area. Some wind machines are likely to be producing power at any given time. However, wind power is not as closely coincident to utility demand peaks as are photovoltaics. Wind machines may generate more electricity in the afternoon in the spring and autumn months.

Collectively, even the most intermittent sources of electricity should qualify for some capacity credit. The California Public Utility Commission staff concluded that even "the nonfirm, surplus-only, short-term contracts with PG&E, providing for what may have appeared to be undependable supply without long-term commitment, have proven to have capacity value."[14] They recommended that small power producers receive a credit based on a 50 percent capacity value until experience justifies some other figure. The full commission overruled the recommendation and awarded intermittent producers 100 percent of the capacity credit firm producers would receive.

Utilities argue that capacity is actually displaced only if the QF meets certain minimum performance criteria. They tend to offer long-term contracts that contain minimum performance criteria. QFs must operate at 70 to 85 percent capacity factors either year-round or during the peak and intermediate-peak periods. Some utilities tried initially to impose a 100 percent capacity factor on QFs, until the commissions pointed out that their own plants rarely operate at better than 80 to 85 percent capacity factors. Often a commission will impose *availability* rather than capacity factor requirements, that is, require that the QF's plant be available for generation for a certain portion of the year or during peak periods. Most utilities do not offer partial payments for QFs that meet partial performance criteria. Some do. Most states give no bonus to a QF that operates more reliably than the utility. Montana is one exception. If the QF exceeds the standard capacity factor needed to gain capacity credits, it will receive a greater capacity payment.

Several states allow utilities to require a "dispatchability" test as part of the performance requirements for capacity credits. New Hampshire's utilities use the same dispatchability requirements as the New England Power Pool (NEPOOL) uses for its members. The QF capacity must be available on two hours notice during the November-through-February peaking period. Thus hydro sites are evaluated to test for minimum flow rates during this period. This is the basis for their capacity credits. New Jersey bases its dispatchability requirements on the Pennsylvania-New Jersey-Maryland (PJM) power pool requirements. PJM also requires its members to meet a two-hour advance notice dispatchability requirement, but, unlike NEPOOL, the PJM requirement must be met at any time during the year. Capacity credits for wind or photovoltaics will be extremely difficult or even impossible to receive under such performance criteria.

Although common, these terms probably undermine the intent of the FERC's requirement that the QF be treated as a class. As Colorado

Hearing Examiner Michael R. Homyak said, "The reliability adjustment calculations will be based upon the characteristics of the aggregate class of the small power producers in question, as it is then constituted, and not upon an individual qualifying facility."[15] The need for minimum performance criteria for individual plants will not be onerous for cogenerators. They have extremely reliable plants over which they have direct control. But it will prove debilitating to small solar electric facilities.

Often utilities try to impose severe penalties on the QF that does not achieve the expected capacity credits. They justify these because actually to displace future capacity, they must include QF capacity in their ten-year resource plan. If, five years down the road, the QF decides to go out of business, the utility will have inadequate capacity. Montana-Dakota Utilities, for example, proposed in August 1981 to pay capacity credits in any month where the QF delivers with at least 65 percent on-peak capacity factor. If the QF fails to meet this criterion, then no capacity payment is made at all and it receives only a nonfirm energy payment. If the QF terminates its service, the "Firm Energy Service" severance penalty payment due the company would be calculated as the average capacity compensation per month for QF power times the remaining months in the unexpired contract. Thus if the QF were on a ten-year contract and earned $100 per month for the first two years and then went out of business, it would have to pay a penalty of $9,600 to the utility! The Montana Public Service Commission disallowed the proposed tariff.

Utilities argue that to gain capacity credits, QFs must displace future generating capacity, and to accomplish this the QF must stay in operation at least as long as the conventional power plant. A QF that stops operations after 5 years cannot be said to displace a coal plant that operates for 30 years. The North Carolina Public Utility Commission explicitly rejected this argument. It noted that PURPA is intended to create a class of dispersed generators. Some QFs will undoubtedly cease operations, but others will enter the market to replace them. Given the ease of entry into this market, if the contractual terms encourage small power producers, utilities should be able to count on continued generation and, therefore, displaced capacity, even if an original producer goes out of business.

Remember, the utility's own power plants do not have 100 percent capacity factors. Nuclear plants operate nearer to 65 percent, and even large coal plants have an effective load-carrying capacity of only 70 to 75 percent of their nameplate rating. Even if we were to use the availability factor rather than the capacity factor, a coal-fired baseload plant does not have higher than an 85 percent factor. The availability factor tells the amount of time the plant is not operating through no conscious design, reflecting the random nature of forced outages. The 85 percent availability factor means 15 percent of the time, or almost five days a month, the plant will not be available to generate electricity even if it is needed for an emergency or to meet a system's peak. The capacity factor is an indicator of the amount of time the plant is actually used, not the amount of time it is available for use. Florida requires capacity credits only to QFs that provide 70 percent "equivalent availability." Montana uses both an 85 percent availability factor and a 70 percent capacity factor in determining its long-term energy and long-term capacity payments.

Small power producers actually displace *more* capacity than their nameplate ratings would indicate. In the language of the grid system, they are located "downstream." Those QFs with capacities under 200 kw will almost certainly be located at the distribution secondary. Thus they avoid transmission and distribution line energy losses. Several utilities have argued before public service commissions that they do not know their exact line losses. Preliminary data indicate that they are usually between 7 and 10 percent. Montana's regulatory commission used 8.3 percent as a first approximation in mid-1982. When questioned by several utilities to justify that figure, the commission conceded it had only preliminary data, but noted that their data provided some evidence of line losses approaching 30 percent in some parts of the state. In its final ruling, the commission requested all state utilities to develop a strong data base for a better estimate of line losses.

Taking line losses into account allows a QF to displace a load greater than its own generating capacity. Thus a 100-kw QF in Montana operating in the distribution secondary would displace, not 100 kw but 109 kw of central plant capacity. In other words, in the absence of the QF, the utility would have to install 109 kw of capacity in a central facility to deliver, after the 8.3 percent line loss, 100 kw at the load.

The line-loss factor is a two-edged sword, however, as it may work against some QFs. Those located in remote sites might have a line-loss percentage assessed against them if they are displacing a peak power facility located nearer the utility's load centers. Wind or hydro facilities might be in this category.

One final point on capacity credits. Many dispersed small power plants can displace more than an equivalent nameplate capacity of central power plants, even though they individually operate at the same capacity factors. As a class, the small power plants will have a greater reliability than the handful of large power plants. In addition, the presence of many dispersed plants can allow the utility to lower the reserve margin it needs. One large plant requires another equally large plant in reserve. Many smaller plants require a lower reserve margin. This increased reliability and decreased need for excess capacity should be taken into account in calculating capacity credits for QFs. This will only be done if they are viewed in the aggregate, as a uniform class, as the FERC's regulations require.

As Available Capacity Credits

Utilities award capacity credits in two ways. One is to require a long-term contract. In this the QF agrees to meet certain performance criteria and to supply a certain minimum amount of capacity. The QF makes the initial decision on how much capacity to commit and also selects the term of the contract. The longer the contract, the greater the capacity credit. Usually the utility estimates the cost of the future power plant and gives the QF a larger and larger portion of the actual kilowatt capital cost as the term of the contract approaches the life of the conventional power plant. Often the terms are linear. Thus, a contract for 5 years will get only one-sixth of the capacity credit paid to a QF willing to sign a 30-year contract.

The second method of disbursing capacity payments is on an as available basis. This method assumes that as a group, QFs will displace a certain amount of capacity. Thus, they should receive a portion of the long-term capacity credit so long as they generate electricity during peak or possibly intermediate-peak hours. The QF is required to meet no performance criteria. The capacity cost in dollars per kilowatt is translated into cents per kilowatt-hour and paid on that basis. As of mid-1982, as available capacity credits have been used in California, North Carolina, Montana, New York, Idaho and Colorado.

Sometimes as available capacity credits are paid only for electricity delivered during peak hours, as in North Carolina. Sometimes, as in New Hampshire, the capacity credits are available on a kilowatt-hour basis at all times. (This case is illustrative of the need for investigating differences in some detail. North Carolina pays, on the average, a penny a kilowatt-hour for electric capacity but only during the peak hours of its peak and off-peak months. New Hampshire pays on average only a half a penny but pays this for all hours of the year. Thus, if North Carolina's peak hours represent 50 percent of the total hours in a year, the capacity credits are identical, depending on the technology's capability of generating on-peak.

Avoided capacity costs are based on the cost of new capacity. The estimation of capacity costs is an arithmetic exercise fraught with controversy. Most states base capacity payments on the annual avoided cost of deferring the addition of a new combined cycle generating unit. To estimate capacity credits, they need to estimate the cost of a new generating unit. One would think this a simple task. They are not, after all, projecting future costs. Power plant list prices are available from manufacturers. But sometimes such a simple task generates varied answers. In 1982 San Diego Gas and Electric estimated a combustion turbine cost of $618 per kilowatt. Pacific Gas and Electric said it cost $777. Southern California Edison estimated $415. The California Public Utility Commission staff bemusedly commented, "We believe that a more consistent value of CT (combustion turbine) cost is called for and request all three utilities to reassess their determination of CT costs and report the results of this reassessment in these proceedings in sufficient detail to allow for comparison between utilities. Reassessment should be based on the value of a generic gas turbine."[16]

The proper value for the capital costs of a combustion turbine in 1982 is probably $350 per kilowatt. For a combined cycle power plant (one that takes the waste steam and uses it to generate more electricity, thereby increasing the efficiency of the plant) it is $650 per kilowatt. A nuclear plant will cost about $3,000 per kilowatt in 1982 dollars and a coal plant will cost $1,800 per kilowatt. The primary difficulty in estimating the costs of these plants is that their planning and construction takes place over several years. Therefore, the time value of money must be taken into account, and the inflation rate for future construction must also be included in the formula. For example, based on the last decade's costs of nuclear power plants, it can be safely predicted that those constructed in the 1980s and 1990s will have cost overruns that exceed the cost of inflation. Therefore, after excluding the impact of inflation, these plants will still cost more (in constant, uninflated dollars) than those being built today. Sometimes utilities underestimate the cost of new plants by not inflating the value of dollars spent on previous years of planning and engineering studies to equate with current dollars. Montana found this to be the case. Its utilities were underestimating the cost of a new baseload plant in this way. They were ignoring the time value of money. A dollar spent in 1973 was worth more than twice a 1983 dollar, assuming an average annual inflation rate of 9 percent. By ignoring the inflation factor, the Montana utilities were underestimating their capacity costs by 20 percent, and therefore underestimating their capacity credits to QFs by 20 percent.

The estimation process is nothing more than a series of guesses. The utility is providing cost data that may or may not be accurate. The New Hampshire Public Service Commission initially investigated the possibility of basing avoided costs on the cost of electricity from the soon-to-be-completed Seabrook Nuclear Power Plant but concluded that the cost estimates were so varied that it could not do so. Final construction costs for nuclear power plants have typically run 100 to 1,000 percent over their initial projections.

Once the flow of future construction investments is developed, the future dollars must be discounted back to their present value. The higher the discount rate (interest, or the cost of future money), the lower the present value of these future dollars and, therefore, the lower the capacity credit. Finally, it must be decided how often these plants will operate (capacity factor) to develop a cost per kilowatt-hour.

For example, New York State in 1982 concluded that Nine Mile Point 2, a 1,080-Mw nuclear power plant, would come on-line in 1986. The cost in 1981 dollars was estimated to be $2.3 billion. Assuming an annual fixed-charge rate of 17 percent to estimate the annual capital costs of the plant and an 80 percent capacity factor, the as available capacity credit for capital alone should be 5.15¢ in 1981 dollars, or almost 6¢ per kilowatt-hour in 1982 dollars. To get this figure, multiply $2.3 billion by the annual fixed charge rate to give the annual cost for the plant of $390 million. The 1,080-Mw plant operates 80 percent of the time, so it produces 7.57 billion kwh annually. Divide the amount of electricity generated into the annual cost of the plant. To this figure, slightly more than a penny in fuel and operating costs would be added. However, nuclear plants historically operate at 55 to 60 percent capacity factors, not 80 percent. Assuming this lower reliability, the capital-related cost would be almost 8¢ per kilowatt-hour in 1982 dollars.

This example illustrates one reason QFs should analyze in some detail the utility's assumptions. For much of the avoided cost data is assumptions. The utility assumes its cost of capital, future inflation rates, future cost of money, the availability and capacity factors of its plants as well as its future demand for electricity. Each of these assumptions must be examined carefully even if they are in the fine print in the appendices of most utility data offerings.

Case Study: Duke Power Company

Capacity credits are sometimes called *demand credits*. They are given for each kilowatt of demand replaced by the QF. Often they are only paid for power generated during the peak periods of the year, requiring a demand meter. Duke Power Company, for example, requires demand metering for customers operating under its Parallel Generation schedule PG. The service is available only to three-phase commercial customers. Peak periods are 7 A.M. to 11 P.M. Monday through Friday. Demand credits are based on the maximum integrated

30-minute demand, which is continuously supplied to the utility during the on-peak periods of June through September and December through March.

A customer desiring demand credits must enter into a contract for a minimum original term of five years. The utility may require a longer original contract term if circumstances indicate this. Thirty months notice of termination in writing is required. Penalties for early termination apply but are not spelled out in the rate schedule. The QF receives $5.12 per month for each kilowatt of capacity if connected to the transmission system and $5.29 per month for each kilowatt of capacity if connected to the distribution system. The annual payment of $60.00 for capacity indicates that a combustion peaking plant ("peaker") was used as the basis for the avoided capacity cost.

Duke Power Company has another tariff for residential as well as commercial customers. It provides for as available capacity credits. The credits vary slightly depending on the contract period. A variable rate contract is offered for a minimum of 5 years, and fixed-rate contracts are offered for 5, 10 and 15 years. Duke Power recognizes it has a secondary peak in the winter, so it provides capacity credits for peak hours during the on-peak months of 1.11¢ per kilowatt-hour and for peak hours during the off-peak months at a lower rate of 0.66¢ per kilowatt-hour. The variable-rate customer will have the rate changed as the fuel charge applicable to retail service is changed. The fixed-rate customer receives no change in the capacity credit.

For this rate class Duke Power also offers an energy credit. The payment varies significantly based on the length of the contract. A QF owner who contracts for the 5-year variable rate receives 2.399¢ per kilowatt-hour for all energy generated on-peak, which is from 7 A.M. to 11 P.M. Monday through Friday, all year. A 5-year fixed contract provides a 2.87¢ per kilowatt-hour payment, and this rises to 5.02¢ per kilowatt-hour for a fixed 15-year contract. The off-peak rate varies from 1.81¢ per kilowatt-hour for a variable-rate, 5-year contract to 3.78¢ for a fixed-rate, 15-year contract.

Case Study: Pacific Gas and Electric

PG&E uses a combined cycle generating plant to determine its cost of capacity. The capacity payment increases with the length of the contract. Table 3–4 presents the annual avoided-capacity costs from plant deferral over 20 years.

Payments are calculated from the schedule of avoided costs as follows: The annual avoided capacity cost for each year of a contract is discounted to its present value. Then the resulting present values are

TABLE 3–4

**Annual Avoided Capacity Costs to PG&E of
Deferring a Combined Cycle Unit Over 20 Years**

YEAR	ANNUAL COST ($/KW/YR)
1980	49.69
1981	53.42
1982	57.42
1983	60.29
1984	63.31
1985	66.47
1986	69.80
1987	73.29
1988	76.95
1989	80.80
1990	84.84
1991	89.08
1992	89.08
1993	98.21
1994	103.12
1995	108.28
1996	113.69
1997	119.38
1998	125.35
1999	131.61

SOURCE: PG&E, *Derivation of PG&E's Full Avoided Costs*, table B, p. 6.

summed and *levelized* (expressed in terms of first-year dollars). The applicable discount rate is 11 percent. For example, consider a five-year contract beginning in 1980. Each of the first five values of table 3–4 is discounted to the 1979 present value (see table 3–5). The present values are summed and then multiplied by the appropriate factor to obtain a levelized payment per kilowatt per year. The result is an annual payment of $56 per kilowatt per year. Note: Discount rates and inflation rates are different. The inflation rate gives one a measure of the rate at which future purchasing power falls. The discount rate is also called the *opportunity cost of money*. It represents the return one might have received from an alternative investment. A great deal of important information can be buried in the footnotes to methodologies. Given that most capacity payments and even firm energy payments are

TABLE 3–5
Calculating Payments from Avoided Capacity Costs

YEAR	AVOIDED COST ($)	×	PRESENT WORTH FACTOR (YR)	=	PRESENT WORTH ($)
1980	49.69		0.9009 (1)		44.77
1981	53.42		0.8116 (2)		43.35
1982	57.42		0.7312 (3)		41.98
1983	60.29		0.6587 (4)		39.71
1984	63.31		0.5935 (5)		37.57

Total present worth = $207.38
Present worth factor (11%, 5 yrs) = 0.27057
Levelized capacity payment = (207.38) (0.27057)
= 56.11, rounded to
56¢/kw/yr

Annual Payments in Dollars per Kilowatt per Year

CONTRACT TERM (YRS)	YEAR OF INITIAL OPERATION					
	1980	1981	1982	1983	1984	1985
1				60	63	66
2			59	62	65	68
3		57	60	63	66	70
4	55	58	61	65	68	71
5	56	60	63	66	69	73
6	57	61	64	67	71	74
7	59	62	65	69	72	76
8	60	63	67	70	74	77
9	61	65	68	71	75	79
10	62	66	69	73	76	80
15	68	72	75	79	83	87
20	73	77	81	85	89	93
30	81	85	89	94	98	103

SOURCE: PG&E, *Power Sales Agreement*, table 1, p. C-5, 4 February 1980.

based on the costs of plants to be built far into the future, the assumed discount and inflation rates can make a very significant difference in the amount of money paid the QF.

Consider, for example, that the inflation rate was 15 percent a year in 1979 and 5 percent in 1982. Assume the QF is bargaining in 1983 and the local utility is building a plant that it estimates will cost $2 billion in 1990 dollars. If the 1990 dollar figure is discounted back to 1983 based on the 1979 inflation rate, it becomes $750 million in 1983 dollars. Based on the 1982 inflation rate, it is $1.42 billion. The higher the present value, the higher the current avoided cost rate will be.

Capacity payments are made only for contracts extending to or commencing from 1983, presumably the date from whence PG&E can defer capacity additions. The prices are subject to change when the state regulatory commission reviews PG&E's capacity costs in future rate applications.

In its Power Sales Agreement, PG&E stipulates a set of minimum performance requirements that must be satisfied by a QF before it qualifies for a capacity payment. The contract capacity for which capacity payments are made is not to exceed the minimum amount of capacity provided during any of the three peak months on PG&E's system (June, July and August). This simply acknowledges that PG&E needs capacity during its peak months. No matter how much capacity is made available during off-peak months, payment will be made only for the minimum amount provided during the peak season. The second criterion is that the contract capacity must be available for 80 percent of the on-peak hours in the three peak months on PG&E's system. The peak hours are 12:30 P.M. to 6:30 P.M. Monday through Friday except for holidays. PG&E, unfortunately, determines the availability factor on an individual or system-by-system basis, when the factor for the entire class of QFs would presumably be much higher.

PG&E defines availability as a percentage of hours, not as a percentage of contract capacity. In other words, the contract capacity may be a fraction of installed capacity that is actually available for 80 percent of the peak hours. For example, a 500-kw cogenerator might decide to commit 100 kw of capacity to be available for 80 percent of the peaking hours, or the owner of a 12-kw hydropower system might commit to a full 12 kw of availability if the stream site were known to be a reliable one.

The contract also indicates that the QF may have scheduled outages other than during the off-peak months only during its first month of operation, presumably to work out any bugs, or when otherwise permitted by PG&E. Capacity payments continue during the scheduled outage period for an annual limit of 35 days, at a daily rate equal to the daily average payment over the preceding month.

If a QF enters into a contract to sell capacity at a future scheduled operation date and the capacity price is revised upward prior to that date, the higher capacity price supersedes the original one.

TABLE 3-6
PG&E's Notice Requirements for Full or Partial
Derating of Contract Capacity

CONTRACT CAPACITY (KW)	LENGTH OF NOTICE (YRS)
Under 25,000	1
25,001 to 50,000	3
50,001 to 100,000	4
Over 100,000	5

PG&E's contract also has penalty provisions. If a QF fails to meet the minimum capacity performance standard, PG&E will suspend capacity payments for a probationary period of up to 14 months. Should the QF demonstrate during that period an ability to meet the minimum obligations, PG&E will make a retroactive capacity payment for the probationary period and reinstate the regular capacity payments at the agreed-upon price. If the QF is unable to meet the minimum requirements during the probationary period, PG&E may either derate (reduce) the contract capacity or terminate capacity payments. Such downward revision in contract capacity will be considered termination without prescribed notice and subjects the QF to certain refund provisions.

PG&E's contract tries to balance the need for the QF to be able to count on future payments with the utility's need to be able to count on a certain future capacity displacement. PG&E requires prior notice in writing for a full or partial derating of contract capacity. Notice has to be given one to five years in advance, depending on the contract capacity. The larger the QF, the longer the prescribed advance notice.

If the QF fails to give advance notice in proportion to its size, it must agree to refund to PG&E the difference between (1) the total capacity payments to date based on the original contract capacity and the current capacity price and (2) the total capacity payments due to the QF over the period of actual previous performance at the adjusted capacity price.

In other words, the QF must pay the difference between what it received under the old contract and the amount it would have collected at the capacity price for as long as it actually performed under the old contract capacity. Interest based on the Bank of America's prime rate is then added. This adjusted capacity price, as PG&E calls it, forms the basis for future capacity payments.

The following example illustrates the result of this capacity derating:

An apartment complex installs a 200-kw, gas-fired cogeneration system to meet its space-heating, air-conditioning and hot water needs. It decides to engage in the simultaneous purchase and sale of electric energy and capacity to PG&E, commencing in 1980. The QF owner(s) enter into a contract capacity for ten years. Five years later, at the end of 1984, the owners notify PG&E that they plan to derate the contract capacity to 100 kw. How much does the QF collect, and how much must be refunded to PG&E?

Original capacity price	$62/kw/yr
Annual payment, $62 × 200	$12,400/yr
Total payment through 1984	$62,000
Adjusted capacity price, 5 yr, 1980	$56/kw/yr
Adjusted contract capacity	100 kw
Adjusted annual payment, $56 × 200 =	$11,200
Annual overpayment, $12,400 − $11,200 =	$1,200/yr
Total refund due PG&E, 5 × $1,200 =	$6,000

Under a complicated formula, the QF that fails to issue prescribed notice must also make a one-time payment to PG&E to cover the excess payments that might have been paid to the QF. For example, a levelized capacity payment might provide for higher payments up front. The utility agrees to this in the expectation of paying lower payments in later years, payments lower than the current capacity credits. If the QF terminates the agreement early, PG&E recaptures the early overpayments.

For example, suppose the same gas cogeneration apartment complex gave 1 month's notice (instead of the prescribed 12) that it wanted to reduce its committed capacity from 200 kw to 100 kw. The one-time penalty payment would be calculated as follows:

Capacity price beginning in 1985 for remaining 5-yr contract	$73/kw/yr
Original contract price	$56/kw/yr
Difference	$17/kw/yr
Times contract capacity, $17 × 200	$3,400
Prorating factor, $1 - \dfrac{1}{12} = \dfrac{11}{12}$	0.9167
Total additional payment, $3,400 × 0.9167	$3,116

PG&E offers the QF three capacity credit options. The QF can choose to receive equal monthly payments. In this case the contracted capacity is multiplied by the contracted price and then divided by 12. The second option is to base the payment on the monthly delivered capacity, which means the payment would fluctuate. Or third, the payment can be made on a cents-per-kilowatt-hour basis for each time-differentiated kilowatt-hour. Thus PG&E offers an as available capacity credit but varies it by the time of day (peak or off-peak value).

Case Study: Montana

Montana has a long-term capacity credit (LTC). The formula for the LTC is simpler than the one for long-term energy (LTE), since the cost of one type of plant need not be subtracted from that of another. The LTC equals $[(b)(d) + f] \times (CF \div 0.85)$. The first part of the equation, $[(b)(d) + f]$, estimates the annual carrying charges of a combustion turbine. CF stands for capacity factor. This is divided by the availability factor of a combustion turbine. It assumes the turbine will be available for service 85 percent of the time. Its random outage rate, or forced outage rate (FOR), is 15 percent. If the QF has a capacity factor of 0.85, then the final equation will be equal to 1. If it has a lower capacity factor than 0.85, the amount of the long-term credit will go down. If it has a higher capacity factor than 0.85, the amount paid will increase. In 1982 the three Montana utilities paid $63.96 to $80.88 per kilowatt per year to QFs signing a minimum four-year contract to deliver capacity.

A QF in Montana willing to sign a minimum four-year contract will thus receive both a long-term energy and long-term capacity payment. The 100-kw, gas-fired cogenerator that operates 85 percent of the time will generate 744,600 kwh per year. It will receive about $38,000 in long-term energy payments and about $6,000 in long-term capacity payments. The 30-kw hydroelectric facility that operates 85 percent of the time will generate 223,380 kwh per year. It will receive $11,400 in long-term energy payments and $1,800 in long-term capacity payments.

This illustrates an important point. Capacity credits do not constitute a very significant portion of the total payments to the QF. However, as will be seen when the economics of small-scale power production are explored, the capacity credits can play a crucial role in changing a deficit into a surplus. They can make the difference between an investment being and not being profitable.

Estimating Base Rate for Firm Energy and Capacity

The following is the formula used by Oregon's Public Service Commission to estimate the base rate for firm energy and capacity. Notice that the associated costs of transmission and operation come to more than $16 million dollars, more than 25 percent of the annual carrying charge for constructing the plant. A 35-year contract gives the QF 100 percent of the firm energy price or firm capacity price. A 5-year contract would give the QF only one-seventh (14 percent) of that price. One could argue that the Oregon formula is quite conservative. It requires an extremely long-term contract before the QF can gain full capacity credits. Moreover, it uses as the cost of a new power plant the Boardman facility, which was already operational in 1982, rather than the cost of the next plant to come on-line.

Computation of Boardman-Based Rates

A = Levelized annual revenue requirement	$62,265,000
B = Coal inventory carrying cost	1,703,000
C = Oil inventory carrying cost	93,000
D = Materials and supplies inventory carrying cost	585,000
E = Operations expense	6,294,000
F = Maintenance expense	5,189,000
G = Transmission expense ($5.56/kw × 424 Mw)	$2,357,440
H = Fuel cost (22.42 mills/kwh)	
n = Number of years in the contract	Varies at the option of the qualifying facility

$$\text{Firm energy price (mills/kwh)} = 0.8577 \frac{(B + C + D + E + F + G)}{424 \times 8{,}760 \times 0.75}$$

$$+ H + \frac{n}{35} \times \frac{A}{424 \times 8{,}760 \times 0.75}$$

This formula reduces to $24.22 + n \times 0.5477$

$$\text{Capacity price (\$/kw/mo)} = 0.1423 \frac{(B + C + D + E + F + G)}{424 \times 8{,}760 \times 0.75}$$

$$+ H + \frac{n}{35} \times \frac{A}{424 \times 8{,}760 \times 0.75} \times \frac{8760 \times 0.75}{12 \times 1000}$$

This formula reduces to $2.20 + n \times 0.04975$

SOURCE: Before the Public Utilities Commissioner, order no. 82–515, 20 July 1982.
NOTES: 0.8577 = Long-run incremental cost-based allocation of total cost to energy
0.1423 = Long-run incremental cost-based allocation of total cost to capacity

TABLE 3-7

NUMBER OF YEARS IN CONTRACT	CAPACITY PAYMENT ($/KW/MO)	FIRM ENERGY PAYMENT (MILLS/KWH)
1	2.25	24.77
2	2.30	25.32
3	2.35	25.87
4	2.40	26.41
5	2.45	26.96
6	2.50	27.51
7	2.55	28.06
8	2.60	28.60
9	2.65	29.15
10	2.70	29.70
11	2.75	30.24
12	2.80	30.79
13	2.85	31.34
14	2.90	31.89
15	2.95	32.44
16	2.99	32.99
17	3.04	33.54
18	3.10	34.08
19	3.14	34.63
20	3.20	35.17
21	3.25	35.72
22	3.30	36.27
23	3.35	36.82
24	3.39	37.37
25	3.44	37.92
26	3.49	38.46
27	3.54	39.01
28	3.59	39.56
29	3.64	40.11
30	3.70	40.65
31	3.74	41.20
32	3.79	41.75
33	3.84	42.29
34	3.89	42.84
35*	3.94	43.39

NOTE: Nonfirm energy is 22.42 mills/kwh. Base standard rate for facilities of 100 kw or less is 35.90 mills/kwh.

* If a qualifying facility enters into a 35-year contract for firm energy and capacity and operates at a 75 percent capacity factor (identical to the design capacity factor of the Boardman plant), the energy and capacity payments combined would equal 50.59 mills/kwh.

The Monopsonist Meets the Monopolist

PURPA creates an internal inconsistency within utilities. On the one hand, the utility is a monopolist who controls all electric sales within its service area. As with any monopoly, its goal is to maximize profits. A by-product of this goal is the tendency to increase rates. On the other hand, PURPA makes the utility the sole buyer of electricity. The monopsonist also tries to maximize profits, but in this case it means minimizing the price it has to pay for power.

The monopolist and the monopsonist should be operating with identical data. Both should be using construction and operating data on current and future power plants. Unfortunately, public service commissions have not yet demanded that both divisions of the utility make their cases at the same proceeding. Therefore, in one hearing the monopolist argues for high rates based on high costs. At another hearing, the monopsonist argues for low buyback rates based on low avoided costs.

For example, utilities treat a consumer of electricity differently from a producer of electricity. The utility does not treat equally a customer's demand for a kilowatt of power and the delivery of a kilowatt of power by the same customer. Utilities will deny avoided capacity payments to power-producing customers on the theory that the utility already had excess capacity, yet will turn around and impose demand charges on these same customers (as power consumers). In 1982 the Niagara Mohawk Power Company proposed to ignore completely the distribution cost savings in its response to a New York Public Service Commission order, but it also advocated a distribution demand charge of between 33¢ and 71¢ per kilowatt whenever the QF's service classification contract demand is exceeded. As Occidental Chemical Corporation (which wanted to sell power to Niagara) argued before the Public Service Commission, "Niagara Mohawk has proposed distribution-related charges for increased demand while not providing a credit for distribution capacity savings resulting from co-generation. This is not justified if one is pursuing an avoided cost-based pricing system." [17]

The commission pointed out that this decentralized power generation is equivalent to not taking power during peak periods. Therefore it concluded "it would be inconsistent to charge back-up and supplementary customers, in particular, for transmission costs when they take service—as Con Edison proposes—but not credit those costs to them when they supply power." [18] The commission in that case concluded "the peak period delivery of power by on-site generators, in and of itself, can enable a utility to avoid transmission capacity costs." [19]

In the above example, Occidental Chemical Corporation argued for a higher avoided cost basis by relying heavily on testimony given by a utility representative in a different rate case. The arguments of the monopolist were given to undermine the arguments of the monopsonist. Based on the data from this other rate case, Occidental recommended avoided capacity costs for cogenerators during peak hours (representing 3,700 hours a year) or 2.35¢ to 4.31¢ per kilowatt-hour. This was four to seven times larger than the avoided capacity credit proposed by Niagara Mohawk in that case.

The war of the numbers continues. Owners of QFs would do well to read transcripts from recent rate cases to gather utility-generated data that could support high buyback rates.

Excess Capacity

The issue of excess capacity is already becoming a central one. Having grossly overestimated future demand, utilities persuaded regulatory commissions to allow them to build several dozen unnecessary plants. Given the 10- to 12-year lead time for large central plants, a decision to build in 1972 when demand was projected to continue to grow at 7 percent a year, doubling every 10 years, would lead to a plant generating electricity beginning in 1984, when demand growth dropped to 1 to 2 percent, or less, a doubling time of 35 to 75 years. Moreover, the economic decline in 1980 to 1983 reduced industrial demand even further. Thus utilities that in the 1970s argued for new plants to meet the needs of the 1980s now argue they need no new electricity for the '80s. But they still envision new plants in the '90s. QFs can legitimately respond that they are indeed displacing new capacity even though that new capacity may not come on-line for more than a decade.

Many utilities argue that they have, or soon will have, a substantial amount of excess capacity. Therefore they argue that they don't need additional QF capacity, at least for several years. The QF should resist that argument. It is a sound one only if these utilities are not planning to build any additional capacity that could be displaced by QFs. That includes any capacity to replace plants to be retired in the near future. Thus, for example, the Potomac Edison Electric Power Company (PEPCO), located in the Washington, D.C. area, argued before local regulatory commissions that it should pay no capacity credit because it has an excess capacity and has no plans to build additional power plants within its ten-year planning horizon. The commissions agreed not to require any capacity credit.

However, during the hearings PEPCO conceded that it was planning to construct a new power plant 11 years hence. By delaying the construction plans by 1 year, it put them outside its 10-year traditional planning horizon, which allowed it to argue that it could displace no future capacity. Understandably, potential QFs were furious. Vigorously, but unsuccessfully, they argued that many utilities could manipulate their resource plans and end up with no capacity payments in exactly the same manner. One way for QFs to respond to these allegations by their utilities is to persuade the city council, rural electric board or state regulatory commission to require the utility to integrate QFs into its 10-year resource planning. PEPCO, for example, estimated no significant QF generation within 10 years. Clearly that was the basis for its plan to construct additional generation capacity in 11 years. However, some commissions, such as in California, have established minimum goals of QF generation capacity that a utility must achieve by a predetermined time period.

The issue of excess capacity is becoming a central one as increasing numbers of utilities defer or cancel new power plants in the light of their overbuilding sprees of the 1970s and declining demand. Ironically, those environmentalists who fought against new power plants are being proven correct. But those plants that were built now represent an excess capacity that is used against small power producers.

However, the QF should remember that if a plant is planned in 1992 or even 1995, it has a value in current dollars. At a real discount rate of 3 percent, capacity worth $50 per kilowatt year in 1995 is still worth $37 per kilowatt year in 1985. The further one delays power plant construction, the lower the capacity credits.

Some public utility commissions not only require utilities to buy independently produced power but to promote it as well. California's public utility commission reduced the return that Pacific Gas and Electric could receive on its investments as a penalty for its lack of effective promotion of cogeneration. California has established quantitative goals for each utility. The commission expects the utility to encourage a specific minimum amount of independent power production before it comes up for another rate increase.

The New Jersey Board of Public Utilities overruled its staff's arguments that QF capacity had zero value because the regional power pool, PJM, had excess capacity. The staff had concluded that a future coal plant would displace oil and therefore no other capacity was necessary. The board found several problems with the staff analysis. First, it was based on quite a few assumptions about costs and inflation rates for the distant future. Second, the only reason a coal plant could be said to displace the need for capacity was because the New Jersey utilities had let themselves become dependent on oil-fired baseload plants. The board noted that "the value of zero for avoided capacity results from the suboptimal fuel mix of the New Jersey utilities. That is, the utilities are presently overly reliant on oil-fired capacity. This, however, will not always be the case. In addition, we are of the opinion that there is intrinsic value to smaller, decentralized cogeneration and small power production facilities." It went on to say that "while the board will not comment on the causes for the present suboptimal fuel mix, we do not want to penalize the development of cogeneration and small power production because of this condition." The board set capacity payments based on the cost of capacity when the electric utility is at an optimal fuel mix. "This results in a capacity value equal to that of a combustion turbine peaking plant." [20]

Given the mistakes in forecasting that most utilities made in the 1970s, the issue of excess capacity will be an increasingly important one in the mid-1980s. The issue has added importance (as well as added complexity) given interstate pooling arrangements. For example, in November 1980, Arkansas Power and Light (AP&L) and the other companies in the Middle South Utilities System entered into a Memorandum of Understanding whereby they would not count cogenerated electricity as capacity on the theory that to do so would jeopardize the massive Middle South construction program. Middle South plans capacity expansion on a systemwide basis. Through a procedure of capacity equalization payments among subsidiaries in four states, every subsidiary pays for its share of new plants regardless of where they are built on the system.

The Little Rock law firm of Nixon and Trotter explained the problem this Memorandum of Understanding presented to QFs in testimony before the Arkansas Public Service Commission. "Before November 1980 any source of capacity brought on-line by a given subsidiary could be counted to determine its capacity equalization payments. Not counting cogeneration capacity has allowed AP&L to argue that avoided capacity costs should not flow from the company to qualified cogenerators since it and Middle South have built excess capacity in the form of coal and nuclear plants to back out (reduce) oil generation." They conclude, "Herein lies the ultimate problem. Plants are not built in Arkansas to serve AP&L ratepayers alone; they are built to serve the Middle South system as a whole. Our experience indicates that the Middle South leadership will not back away from its construction program. Consequently few should be surprised at AP&L's cogeneration strategy." [21]

This problem is difficult and formidable. One of the problems is that state commissions have little or no authority over regional power pools. Indeed, in 1983 Middle South Utilities requested permission of the FERC to set its own PURPA rates, bypassing the state commissions. Independent power producers and the regulatory commissions protested against this usurpation of state authority. The FERC had not yet decided the issue as this book went to press, but it raises a delicate issue, and one that will continue to plague the potential power producer. Only if the FERC imposes the same tough guidelines on new capacity for regional power pools as state commissions have recently done for in-state systems can this problem be overcome. Yet the FERC under Reagan has been predisposed to favor utility interests.

The problem is prevalent throughout the nation. Southern California utilities, for example, plan to get future base load from coal-fired plants in Utah, Arizona and New Mexico. California utilities will own only a few peaking plants. Does this mean that when such construction programs are completed, the Utah, Arizona and New Mexico utilities will argue before their commissions that they have excess capacity and therefore should pay no capacity credits?

The relationship of regional power pools and joint ventures to a local utility's avoided costs is already proving crucial to capacity credits. Sometimes utilities use regional power pools in one way during PURPA proceedings and in another way when they are involved in rate cases. For example, in one rate case PEPCO argued that it could not rely on the PJM power pool for power, that it had to build its own power plants to meet future capacity. In the PURPA proceedings, on the other hand, it argued it should pay no capacity credits because PJM was building sufficient capacity to meet all its future needs. The argument is more difficult to unravel when there is a joint venture. Municipal utilities, for example, might own a small share in a coal or nuclear plant scheduled to come on-line in five or ten years. Generation and transmission (G&T) cooperatives sell shares to distribution cooperatives. In these cases the local utility could argue that it signed a letter of intent to buy into future capacity and cannot withdraw that intent even if QFs displace the need for such capacity. Regulatory commissions are taking an increasingly aggressive position with respect to this attitude. They may decide that the decision to buy future capacity was a management mistake based on erroneous load forecasting (QFs can be considered as load displacement the same as energy conservation). Therefore it could ask the utility to sell its share in the power plant. Montana's Public Service Commission has already warned its utilities that the commission's responsibility is to assure cost-effective power to the ratepayers, and if the utilities are recalcitrant in encouraging QFs, they may be viewed as irresponsible and the rate of return on their rate base can be decreased, thereby reducing their profits.

The Montana Public Service Commissioners are directly elected. Each commissioner represents a different part of the state. As a result, the commissioners are more responsive to the popular will, which brings us once again to the importance of politics in PURPA proceedings. Again and again those in favor of independent power production have successfully used politics to strengthen their bargaining position. In Indiana the state legislature overturned a public service commission's pro-utility PURPA decisions. In North Carolina the state legislature set minimum buyback rates. In Oregon the state legislature required the utility commissioner to develop a system that gave the highest possible buyback rate to independent producers despite the very low price for electricity paid by publicly owned systems.

Political pressure means organization. Even in California, where one might expect a sympathetic regulatory commission and state legislature, small power producers concluded in early 1982 that they could protect their interests only if they organized an independent coalition that could continually lobby.

When Utility Purchases Are Not Required

PURPA 210 allows utilities to stop buying energy or capacity from QFs whenever doing so would increase their operating or production costs. To many observers this section of the law is a potential escape hatch. The FERC requires utilities to give advance notice of such times and to prove to the satisfaction of the regulatory commission that its avoided costs are actually negative during that time.

The major problem with this section of the law is that it may actually make QFs vulnerable to the poor planning of utilities. For example, a utility's current avoided energy and capacity costs might be quite high, but it plans to bring a nuclear plant on-line in two or three years. That plant is expensive and also unnecessary because demand has softened to the point that added capacity is no longer needed in the projected time frame. However, once it comes on-line, it may be less expensive to run it than to leave it idle. As previously shown the fuel costs of a nuclear plant are the lowest of any steam electric plant. Nuclear plants must be operated at a certain load. If they dip below a certain load they must be shut down. Once shut down it takes a long time to start them up again. Therefore, the utility can argue that to purchase electricity from QFs when the nuclear plant is operating at partial load could displace sufficient energy to force a shut-down of the nuclear plant. The costs of shut-down and start-up would actually raise the overall operating costs. Therefore the utility need not buy QF energy during this time.

Most regulatory commissions specifically cite such a case as one that would allow utilities to cease purchases from QFs. The problem is that by allowing this, the regulatory commission makes the QF pay for planning errors by the utility that have led to excess capacity.

The best response by QFs is to ally themselves with those vigorously promoting conservation to fight against any expansion of central power plants. In several states, such as Kansas and Pennsylvania, the regulatory commissioners ruled that stockholders had to pay for cost overruns of nuclear plants and capacity that comes on-line but is unneeded. The more the costs of poor planning are imposed on stockholders rather than ratepayers, the greater will be the tendency of the utility to forecast cautiously. The more that happens, the less the QFs will be faced with a fait accompli, that is, an excess capacity due to previous poor projections that leads the utility to reduce its purchases of QF power.

In 1981 PG&E submitted a preliminary Power Sales Agreement to the California Public Utility Commission. The agreement contains the following section entitled "Continuity of Service."

> *PG&E shall not be obligated to accept, and may require Seller to curtail, interrupt or reduce deliveries of energy or energy and capacity (1) whenever PG&E can obtain energy from another source, other than a PG&E fossil-fueled plant, at a cost less than the price paid to Seller, (2) during any period when PG&E can generate or purchase an equivalent replacement amount of electric energy generated from renewable resources . . . or from plants designated for operation to minimize air pollution, or (3) during periods of minimum system operations.*

This contract says that PG&E could refuse all QF purchases whenever actual avoided costs fall below the stated buyback rate. This confuses the FERC directive that says utilities may refuse to accept QF output when such acceptance in itself would raise the overall cost of service to its customers. The FERC's intention was *not* to allow utilities to refuse purchases just because they miscalculated avoided costs prior to QF delivery. The FERC, for example, allows the QF the option of selling energy to the utility on an as available basis, the price for which is to equal the avoided costs at the time of delivery. But it also allows the QF to elect to sell energy on the basis of a contract that estimates future avoided costs. In the latter case, the utility must continue to pay the QF on the basis of the costs estimated when the obligation was made, even if avoided costs actually drop below that level.

Fortunately, PG&E qualifies the above provision with another that says, "PG&E shall take or be prepared to take energy or energy and capacity from Seller for not less than 8,160 hours of each calendar year." Thus PG&E has restricted its authority to invoke the no-purchase clause to about 7 percent of the year.

Most regulatory commissions have recognized that this provision of PURPA could be used as an escape hatch for utilities that are reluctant to buy from QFs. They strongly urge utilities not to do so, and they require utilities to present data to support their contention that negative avoided costs are present during those times. They also require advance notice. Nevertheless, the only guaranteed protection is a provision, as in the PG&E contract, that the no-purchase allowance is permissible for a limited number of hours a year.

All-Requirements Utilities and Wheeling

All-requirements utilities are those that buy all their power from other utilities. They often have what are called "take or pay" contracts, which say they must pay for a certain amount of electricity whether or not they use that amount. Also, they must buy all their electricity from a certain supplier. These utilities may try to use this contract as the basis for refusing to buy power from a QF. The FERC specifically notes however, that these all-requirements utilities have the same obligation to buy as do other utilities. PURPA Section 210 supersedes any existing take-or-pay contract. Thus the local utility must buy QF energy, which then displaces the purchase of energy from its wholesale supplier.

The owner of a QF should keep in mind that the avoided cost to the all-requirements utility is nothing more than its purchase price of electricity and capacity from the bulk supplier. This purchase price is based on embedded or historic costs, rather than marginal costs. Many all-requirements utilities purchase wholesale power for only a couple of cents a kilowatt-hour. Thus, for example, the Clark County, Washington, Public Utility District (PUD) was offering QFs their average system retail rate of 2¢ per kilowatt-hour in late 1981. At the same time Bonneville Power Administration (BPA), the supplier to Clark County, calculated its long-run incremental cost for energy at 6.18¢ per kilowatt-hour and the value of capacity at $55.32 per kilowatt. The advantages of selling to the supplier rather than to the local utility in this case are obvious.

The all-requirements utility has the right, subject to the approval of the QF, to wheel or transmit the QF's electricity to the bulk supplier. The bulk supplier must then purchase capacity and energy as if it were the original buyer. It must pay a price based on its cost of avoided power. Such a price should be much greater than its wholesale price to the all-requirements utility.

Most all-requirements utilities are publicly owned. Thus QFs can organize to influence decisions about wheeling. They can participate before city councils or in rural electric cooperative membership meetings. Or they can lobby the state legislature to direct the utilities to wheel the electricity to better-paying, more remote customers. This last strategy was successfully pursued in Oregon.

The Oregon legislature took an innovative approach to the profound difference between the avoided cost of power to its public utilities, which are largely customers of the BPA, and the avoided cost of power to BPA itself. In 1981 the legislature amended the law to give the Public Utility Commissioner jurisdiction over municipalities, cooperatives and people's utility districts for the purpose of PURPA Sections 201 and 210. The commissioner then ordered all utilities in Oregon to pay a QF its avoided cost of power or a base rate, whichever was higher. The commissioner noted that the base rate's "primary purpose was not to establish a new rate for electricity which was above any utility's avoided cost, but rather to provide an incentive for certain utilities to wheel power voluntarily, rather than pay a rate which exceeded their own avoided cost. Such a device was thought necessary by the Legislature because of a presumed inability of the state to mandate wheeling, in the light of the possibility of federal preemption [because wheeling is considered to be interstate commerce, under federal jurisdiction]."[22] Moreover, the establishment of a base rate would give prospective QFs a better idea of the minimum payment they could receive under the contract.

Oregon's investor-owned utilities had avoided costs of up to 6¢ per kilowatt-hour in 1982 for a long-term contract and 3¢ or more per kilowatt-hour for short-term firm power. The BPA maximum wholesale rate to preference customers at the time was 0.74¢ per kilowatt-hour for energy and $2.80 per kilowatt of billing demand. The Oregon commissioner established a base standard rate for facilities of 100 kw or less by using the lowest standard rate available to facilities interconnected with investor-owned utilities in Oregon. That rate was 3.59¢ per kilowatt-hour, which was adopted as the base standard rate.

For those wanting a variable rate for long-term contracts, the commissioner used as a basis the highest-cost permanent baseload plant serving Oregon consumers. It was an existing coal-fired electric plant owned by Portland General Electric (PGE). He assumed the cost of a publicly-owned plant would be only two-thirds that of the PGE plant due to the lower financing costs of publicly-owned facilities. He then established a table to estimate the base rate for varying contract terms. A 12-year contract at a 75 percent capacity factor would give 3.59¢ per kilowatt-hour, the same as the base standard rate. QFs who signed longer contracts received a higher rate.

The Problem of Falling Avoided Costs

The price of electricity can be expected to rise for the foreseeable future, but this does not necessarily apply to the price QFs will receive for their electricity. Avoided costs could decline for several reasons. California's utilities switched from expensive oil to less expensive natural gas for their peaking and intermediate peak plants in 1982. Their fuel costs dropped, and as a result they dramatically reduced the price they offered to QFs. This is a temporary situation because natural gas deregulation will raise its price to parity with oil by mid-1985. But this did not stop at least one QF from going bankrupt. He had counted on the 7¢ per kilowatt-hour price in 1981 and couldn't repay his investment with the offered 4.5¢ of 1982.

The declining price of oil between 1981 and 1983 reduced buy-back rates around the nation and emphasized the risky nature of small power production. PURPA guarantees the QF a market; but it does not guarantee a profit. The utility company is guaranteed a return on its investment, but the QF is at the mercy of external forces. Anyone entering the field would do well to carefully analyze future trends in electric demand and fuel prices.

A more permanent drop in avoided costs will take place when the top of the load curve is displaced by conservation or QF generation. Say, for example, that the top (peak) of the load curve is carried by 200 Mw of expensive, inefficient, oil-fired power plants. After 200 Mw of peak capacity is introduced by QFs or when 200 Mw of peak capacity is shaved through load leveling or other conservation efforts, this inefficient, high-cost plant will be displaced. As the peaks in the load curve get lower, less expensive plants will carry peak load, and avoided costs will drop.

Some states recognized this problem very early. They wanted to assure the first QFs coming on-line that they would not be penalized later when more QFs came on. That is, those who took the risk of being first should receive the highest return. Those who came five or six years afterward would receive a lower avoided cost rate, which would not be applied to the first QFs. New Hampshire's public service commission actually enacted a "grandfather" clause, which ordered that the purchase price of QF energy never drop below the initially-contracted avoided cost price. If avoided costs rose, the purchase price would also rise. If they dropped, the purchase price would remain at its original level. However, the New Hampshire State Supreme Court overturned this regulation. The court ruled that by enacting such a regulation, the commission was in effect restricting the authority of future commissions. This represented an inappropriate exercise of its own power.

New Jersey's Board of Public Utilities may be more successful in insulating its QFs from potentially declining avoided costs. It ordered utilities to pay an energy rate equal to 110 percent of the projected annual average *running rate* of the PJM. The running rate represents the average hourly cost of the marginal fuel plus variable operation and maintenance costs of dispatching PJM plants to serve an additional kilowatt-hour of demand. In 1982, on an individual hourly basis, this value ranges from a penny a kilowatt-hour for a coal plant to 11¢ a kilowatt-hour for a gas-fired jet turbine. The board selected this measure as the best estimate of electric utilities' avoided energy costs. It believed the estimate to be conservative and also wanted to encourage QF development, so it added the 10 percent premium.

In New Jersey the energy rate is subject to retroactive upward adjustment if the actual rates are higher than the projected rates. No downward adjustment is possible even where the projected avoided energy costs overstate actual costs. In 1980 the actual average running rate was 5.30¢ per kilowatt-hour. Keep in mind that, because of inflation, any long-term fixed-price contracts actually yield a declining real value of the money paid. A nickel ten years in the future will be worth considerably less than today's nickel. On the other hand, a long-term contract protects the QF against a dramatic decline in avoided costs, as occurred in California in the spring of 1982.

Service or Customer Charges to QFs

PURPA requires that QFs be charged no higher rates for back-up service, maintenance power, standby power or interruptible power than customers that have no generating capacity, unless the utility can prove it has a higher cost of serving QFs. Most regulatory commissions put the burden of proof on the utilities to document the increased cost of service.

However, most utilities do impose unique customer charges on QFs. Some are for repaying extra metering costs, and these are appropriate so long as the actual cost of the meter is based on the cost to the utility for bulk purchases and not the cost to the QF for purchasing a single meter. Thus Union Electric in Missouri in late 1981 required a $4-per-month charge for a single-phase energy meter and $6 for a three-phase energy meter. If the customer wanted to be paid for time-differentiated energy (peak and off-peak), the meter cost $13 per month for single-phase and $15 for three-phase.

Some utilities charge stiff customer service fees. Sulphur Springs Valley Electric Cooperative in Arizona, for example, charges $6.50 per month per meter for handling the QF's account. PEPCO charges $11 per month unless the QF's electricity is delivered at 13.2 kilovolt (kv). If it is, the customer charge becomes $80 per month and there is a reduced energy payment. The reduced energy payment is presumably a result of the QF delivering the power to the primary side of the distribution transformer. The QF avoids the line losses inherent in the transformer and thus receives a lower payment because the electricity will probably be consumed on the secondary side of the transformer. Several New York utilities submitted tariffs to the public service commission in mid-1982 that required $100 monthly customer charges for QFs. The QF can complain to the regulatory commission and ask that it request documentation from the utilities to show that serving a QF is so much more costly than serving a regular customer.

Some utilities have tried to charge higher rates to QFs for back-up power. This is conceivable when the QF operates so that it uses power internally and only buys from the utility irregularly. However, even in this case the FERC requires that utilities not assume that such down-times will be coincident with their system peaks. For QFs that buy all their electricity from the utility and sell all they generate to the utility, discriminatory back-up charges should be easily denied by the regulatory commission.

Wheeling

Wheeling is defined in general terms as the transmission of power to a purchaser on behalf of a generating entity by a third party where the transmission facilities are owned by the third party.

The ability to wheel electricity across the nearest utility's lines to a remote utility opens up larger markets for the QFs power. PURPA allows, but doesn't require, wheeling if the utility and the QF agree to it. The FERC comments, "the commission notes that this transmission can only occur with the consent of the utility to which energy or capacity from the qualifying facility is made available. Thus no utility is forced to wheel."[23] The United States Supreme Court has concluded that the Federal Energy Regulatory Commission cannot order wheeling except when they can prove that the refusal of the utility to wheel electricity is part of a larger restraint of trade under the antitrust doctrine.[24] In the 1930s Congress debated including a "common carrier" provision making it the duty of every public utility to transmit energy for any person upon reasonable request. This is already the case for oil pipelines and railroads. These provisions were eliminated to preserve "the voluntary action of the utilities."[25] The Fifth Circuit Court concluded in 1981 that "the legislative history of the Federal Power Act makes clear that the commission lacks the authority to require electric utilities to provide wheeling even upon a reasonable request."[26]

A state's authority to order wheeling is circumscribed. Federal courts have held that states were preempted under Section 201 of the Federal Power Act from requiring the transmission of electric power and the sale of wholesale electric power in interstate commerce.[27] However, the wheeling issue may become a central one in the PURPA debate. As noted above, the Oregon legislature designed an innovative base rate structure designed to encourage public utilities to wheel electricity to investor-owned utilities or regional suppliers who would have a higher avoided cost rate. Two-thirds of all municipal utilities have no generating capacity. Their avoided costs are therefore based on their contracts with bulk suppliers. If the municipal utility purchases a QF's power, it pays that wholesale cost. The difference between that and the avoided cost to the bulk supplier may be two or even three to one.

New Hampshire's Limited Electrical Energy Producers Act of 1979 requires wheeling. To date it has not been tested. Several New Hampshire utilities have formally expressed their willingness to wheel any QF power to a remote supplier. California's Public Utility Commission noted in a decision in late 1980 that "we agree that wheeling is an important concept that can make a material contribution to cogeneration and small power production in California . . . this commission will view with disfavor any actions or inactions by a utility which will interfere with the signing of a contract between a neighboring utility and a cogenerator or small power producer located in its service territory."[28] Minnesota enacted legislation that requires, for all QFs with capacities less than 30 kw, that the utility "at the QF or the utility's request, provide wheeling or exchange agreements wherever practicable to sell the qualifying facility's output to any other Minnesota utility having generation expansion anticipated or planned for the ensuing ten years."[29]

Little investigation has occurred on appropriate wheeling charges. Testimony from two witnesses before the Board of Public Utilities in New Jersey indicates the differences of opinion. James Donald Hebson, Jr., manager of transmission planning of the Public Service Electric and

Gas Company (PSE&G), indicated that his methodology for estimating wheeling costs would be to set all of PSE&G's "loads and generator outputs . . . equal to zero."[30] He would treat the system as if it were virgin and estimate the costs of new electricity coming on-line. This would be terrible news to the independent power producer. Edward P. Kahn, a mathematician who has testified before many regulatory commissions, addressed the same issue by stating, "Since there are currently very substantial power flows in New Jersey relative to the size of QFs . . . it is a reasonable assumption that wheeling will impose no new costs and hence QFs should not be charged for this service."[31]

It is by now apparent that issues such as avoided costs, short- and long-term marginal rates, cost of peaking plants, length of contract, levelized payments and the like are complex. The reader is advised that they are also quite new. In the future the feedback from this first round of contracts and rates will generate a more sophisticated approach, and probably a simpler one. As more data is gathered, the relationship between rates and investments and the reliability data on solar electric systems will become clearer.

In this chapter the contractual price has been the primary focus. Chapter 4 discusses an equally important and complex topic: interconnection standards. How much does it cost to interconnect with the utility, and how can the quality of electricity sent into the system be maintained at a sufficiently high level?

CHAPTER 4

Interconnecting with the Grid

> Once you interconnect a generator and operate it in parallel with the utility network, it becomes part of a vast sophisticated machine. So it needs sophisticated protection; protection against what the network can do to the generator, and protection against what the generator can do to the network. And right now, no one, including your utility, knows how much protection is prudent and how much is overkill, in your specific interconnection situation.
>
> *Power* magazine, June 1982

The Public Utility Regulatory Policies Act (PURPA) gives the qualifying facility (QF) the right to interconnect with the electric grid system. It gives the utility the right and the responsibility to impose interconnection standards necessary to protect its workers, the equipment of its other customers and the overall quality of electricity in the system.[1]

Utilities are used to dealing with other utilities and to linking together a few dozen billion-dollar power plants. Under PURPA, utilities will have to interconnect with hundreds of thousands, possibly millions of power plants, some costing no more than $2,000 to $5,000. Standards relevant to central power plants that serve thousands of customers may not be relevant to thousands of dispersed power plants.

Meanwhile, as more and more dispersed plants come on-line, a demand for low-cost, high-quality interconnection equipment is created. Already the electronics industry is introducing products to meet this demand. Standards are thus being developed even as the technologies are rapidly evolving.

No one knows just what type and level of protection is required; yet the costs of interconnection can vary by an order of magnitude (ten to one) depending on the standards imposed. For small power producers, especially those with a capacity less than 50 kilowatts (kw), the level of the standard can spell the difference between a profitable and unprofitable facility.

Even in an ideal situation with mutual respect and understanding prevailing, the utility and the QF approach the interconnection issue differently. The utility engineer plans for the worst case—the worst combination of theoretical circumstances that could harm the system. In doing this he errs on the side of caution. The QF, on the other hand, wants to maximize its revenue and errs on the side of aggressiveness.

Often these different approaches breed mistrust and suspicion. Lacking the electrical expertise and caution of utility engineers, the QF may believe the very sophisticated standards required by the utilities are nothing but a ploy to discourage independent power produc-

Photo 4–1: The basic components for interconnecting a wind system to a utility are the wind machine control panel (the box at the left) and the synchronous inverter (the second box from the left). The other boxes are for data collection for this particular installation. *Photograph courtesy of Joe Carter.*

tion—and sometimes they are. On the other hand, worried about the possibility of legions of small generators owned by amateurs primarily motivated by visions of instant wealth, utility engineers may view the entire category of small power producers as a nuisance and a danger.

Compounding this problem is the fact that no interconnection standards are yet in effect. Some national organizations, like the National Rural Electric Cooperative Association (NRECA) and the American Public Power Association, have formulated general guidelines. The International Electronic and Electrical Engineers are developing draft

Photo 4–2: Shown here is a dual metering system. Reverse flow equals total wind system output to the grid (surplus overload), and the billing meter equals the total house load minus any wind system input to the house load. *Photograph courtesy of Joe Carter.*

standards. Several utilities have also developed general guidelines. But in most situations, interconnection policies are developed on a case-by-case basis. The customer submits detailed plans to the utility, and the utility's engineers approve, revise or disapprove these plans.

Moreover, public service commissions have tended to stay out of the interconnection regulatory area. They leave this up to the individual utilities. Some commissions, such as Michigan's, have established investigating commissions with QFs on the advisory board to assist in the setting of interconnection standards. Most commissions prefer to wait for sufficient data about possible safety and protection problems before making final decisions.

PURPA makes little mention of interconnection except to note that the QF must pay all reasonable costs. These include costs of connection, switching, metering, telemetering, dispatching, transmission, distribution, safety provisions and administrative costs incurred by the electric utility that are directly related to the installation and maintenance of the physical facilities necessary to permit interconnected operations with a QF. However, PURPA regulations limit the amount the QF must pay to that portion of the interconnection costs in *excess* of the corresponding costs the electric utility would have incurred if it had not engaged in interconnected operations, but instead, had generated an equivalent amount of energy itself.

Protecting the Worker, the Customer and the Network

When a large number of small, dispersed generators are added to the grid system, three possible problems arise. One concerns the safety of people working on the line. The second relates to the quality of electricity in the system and the impact of the dispersed power plant on equipment owned by the QF and other customers. The third concerns the additional costs that QFs might impose on the grid system, for example, if the QF's equipment draws a large amount of reactive power (explained later in this chapter) from the grid.

The problem of worker safety is relatively straightforward. Utility personnel working on the transmission and distribution (T&D) system face the danger of receiving a shock or even of being electrocuted if a line or a piece of equipment undergoing service suddenly becomes energized by a cogenerator or small power producer. To resolve this problem, all utilities require the QF to install disconnect switches that can be locked in an open position while the workers are servicing that part of the line. The major controversy about such switches is where they are to be located and whether in remote locations they can be controlled ·automatically by utility personnel. Linesmen may spend little time manually disconnecting one or two generators, but if the penetration of small power plants reaches a large level, manual disconnection will take quite a number of hours. On the other hand, if the utility can disconnect from its central headquarters, the QF may relinquish control over the generator.

The problem of equipment protection and maintaining the overall quality of electricity in the system is a more difficult one. Appliances and other equipment are designed to operate with alternating current of a certain quality. Equipment damage is likely to occur if the voltage and frequency deviate too much from certain values. In addition, all types of electrical equipment will be damaged if supplied with excessive current.

Utilities try to protect against these distortions by installing relays and circuit breakers, and they may require over/under current relays, over/under voltage relays or over/under frequency relays. These disconnect the QF automatically when the current or voltage of frequency vary by more than a predetermined amount. Relays vary significantly in price depending on their reliability and the range within which they operate.

The small wind machine or cogeneration owners may find it hard to believe that anything their generators do can affect the utility system. In fact, their generators will not affect the entire system, but rather those customers with whom they share a transformer. Five to 20 customers may share the same transformer. If the generator is large enough, it could even affect larger numbers of customers that share the primary feeder.

For most people electricity is a rather abstract commodity. It is hard to imagine an electron, let alone the wave form of voltage or current. Yet these insubstantial characteristics of electricity play an important role in operating the machines around us. For example, variations in the frequency from the basic 60 cycles per second (60 hertz) can change the way modern appliances operate. Timers in the kitchen range, microwave oven or clock use this power frequency. If the frequency were allowed to err by only 1 hertz (Hz), that is, 1 cycle per second, the electric clock would gain or lose 1 minute per hour or 24 minutes per day. The TV may be afflicted with fluttering or creeping ghosts. The stereo may play sharp or flat. Indeed, many appliances simply will not operate on abnormal frequencies. They contain circuits, like power factor correction or filtering circuits, that rapidly unbalance as a function of power frequency. Unless the appliance carries the legend 115 v/50–65 Hz or something similar, it may not function with "off frequency" power.

Significant voltage fluctuations can also damage equipment. Utilities normally provide their customers with power within 5 percent of the nominal voltage level. The most common type of electric service is 120/240 volts (v), single phase. The acceptable limits for this particular type of service are 114 to 126/228 to 252 v. Very few customers will experience the full 12-v to 24-v variation. Typically, customers near the distribution substation will receive the top end of the range, while those at the end of the service line will receive voltages around the middle or low end. In order to maintain proper voltage levels, utilities install voltage regulators throughout the T&D system.

Voltage regulation could become a problem if dispersed power generators with highly fluctuating outputs come on-line. When considered per unit of dispersed generator capacity, technologies such as wind turbines and photovoltaic systems present a greater voltage regulation problem than conventional level-running cogeneration systems. One study assessed the voltage response in a distribution feeder when many small wind turbines were introduced.[2] This study is important because the highly fluctuating nature of outputs from wind turbines may represent the worst-case scenario for such interconnection problems as voltage regulation. The study concluded that at wind turbine (market) penetration levels below 20 percent only minimal changes in the voltage profile throughout the feeder were observed. Even with 50 percent penetration, the voltage levels in the feeder generally remained within the plus or minus 5 percent utility norm.

Voltage flicker is another common problem. For customers in rural areas, voltage flicker is often the norm. The surge of current to an induction machine during start-up can cause a fluctuation in the voltage in the rest of the utility's distribution system. In 120-v residential systems, a sudden voltage change of 6 v to 8 v or more will result in objectionable light flicker. For example, light flicker is common in homes when the refrigerator turns on and off. Induction generators in particular can cause voltage flicker, since they draw a large inrush current for a few cycles while starting. Voltage flicker problems are most likely to arise in a distribution secondary if a small induction generator shares a distribution transformer with other customers.

Most utilities have general guidelines pertaining to voltage flicker, but they have not tended to adopt requirements related to voltage flicker when establishing interconnection policies under PURPA.

Faults and Reclosers

Fluctuations in current levels often result from faults in the electrical system. A fault is any failure that interferes with the normal flow of current. There are two types of faults: shorts and opens. *Shorts,* or short circuits, occur when there is an unintentional connection between parts of the circuit, which results in high current flow in the supply conductor. The undesired connection of one or more lines to ground results in unbalance, and is called a *ground fault*. *Opens* occur when the circuit is broken; the result is that no current flows. Faults in the electrical transmission and distribution system are caused by occurrences such as equipment malfunction, adverse weather conditions and tree limbs contacting lines.

To prevent equipment damage, utilities use protective devices to isolate faulted sections. These devices are located throughout the T&D system. They contain the fault to as small a section as possible. Since many faults are of a transient nature, fault protection devices include devices called *reclosers*. These open and close the faulted line a set number of times before permanently disconnecting it if the fault persists. Automatic reclosers can close and open as many as five times until the fault clears. After that the recloser locks out, isolating the line for manual intervention.

Most often manual intervention isn't necessary, and the customer experiences only a modest voltage flicker. But even the short interval the automatic recloser cuts in and out can have a damaging effect on the dispersed generator. One possible problem is that the recloser could connect two live generating systems at a moment that the voltages and frequencies of the dispersed generator and the utility's generators are not exactly in phase. If they're not, the utility system, with the inertia of several large generators behind it, can jerk the generator into phase in a fraction of a second. "If all you get is a sheared generator shaft, consider yourself lucky," *Power* magazine advises.[3]

Fortunately, induction generators and line-commutated inverters (see below for description) will normally shut down much faster than the reclose interval when utility power is lost. Furthermore, reclosers are in some cases equipped with a voltage relay that prevents them from reclosing on an energized line. The utility is just as interested as the QF that it not remain connected when the reclosing circuit breaker is attempting to clear the fault. The QF could continue to feed the fault, keeping the arc alive.

The dispersed generator must be in phase with the utility generators. If they are out of phase, the power flow from the combined or interconnected systems can cause damage to electrical machinery. Dispersed generators must be equipped with a device that properly synchronizes the electrical wave patterns that are produced with alternating current.

Synchronization is useful even if the small power producer is not interconnected with the grid but is operating two machines in synchronization in a load-sharing manner. For example, a home or commercial establishment might have a hydro plant and a wind turbine. When the wind is blowing strongly, the hydro could be shut down entirely, and vice versa. This would conserve water in the impoundment, an important consideration in most hydro installations.

Protective Relays

To protect other customers and maintain system quality, relays are needed. The purpose of protective relays is to detect unsafe or out-of-limit conditions in a power system and to trip appropriate circuit breakers. The prime intent of the protective relay circuit is to isolate the faulty section of the circuit so that the remainder of the network can continue to deliver power without interruption.

Relays are a relatively esoteric topic even for professional engineers. *Power* magazine, a journal for professional engineers, admits that even "electrical engineers usually have little more than a casual understanding of protective relaying, and even most power plant engineers are only dimly aware of its existence. It's a foreign subject, talked and written about with foreign symbols that the average power user doesn't (usually) need to understand. . . . But for the utility engineer, protective relaying is a way of life. The utility's relay engineers spend all day figuring out possible combinations of faults that might occur in the network, and designing relay circuits to protect against them. In the process, they have developed a highly systematic approach to relaying and an efficient shorthand for designing circuits and specifying relays. . . . Most small customer/generators know nothing about relay engineering, and the utility engineer is faced with the job of teaching them. It's a lesson that the customer doesn't want to learn and the relay engineer doesn't want to teach."[4]

Utilities usually demand the highest quality relays, appropriately called *utility-grade relays*. These devices are very reliable and should last for 40 years. They have a relay trip setting that can be adjusted accurately to known values. Each is equipped with indicators showing that it has tripped and why it has tripped (over or under voltage, for example). The insides of utility-grade relays can be removed for servicing.

Another common type of relay is an *industrial-grade relay*, often found inside factories. This kind of relay is much less reliable than a utility-grade relay and lasts for only four to five years. They are also about a third less expensive than utility-grade relays. Some utilities, such as Pacific Gas and Electric Company (PG&E), permit the use of industrial-grade relays with small and medium-size dispersed generators (less than 1,000 kw). The company does not feel that the higher reliability and accuracy of utility-grade relays are necessary for these applications. On the other hand, one hydro developer in Pennsylvania who initially fought against Pennsylvania Power and Light's requirement that utility-grade relays be purchased later changed his mind and believed that the longer life and higher reliability actually made them a better purchase, even for his relatively small facility.

After the relays are installed, the next issue is what their trip settings should be. Standard relay settings have not been established. Typical settings are plus or minus two cycles per second on frequency and plus or minus 10 to 20 percent on voltage. Relay settings are a source of misunderstanding between the utility and the dispersed generator. The utility's standard approach is to drop a generator off-line the moment trouble is detected. The QF, on the other hand, wants uninterrupted power for his or her household or business, or wants uninterrupted revenue from exporting the electricity.

A typical problem for industrial cogenerators occurs when the voltage surges on the utility's line as a result of switching of banks of power-factor correction capacitors (see the discussion of power factor below). Often these surges are sufficient to trip the under-voltage relays and drop out the customer's generator. The utility doesn't give any notice about these surges. According to *Power* magazine, "Right now, customers that are interconnected to relatively small utility feeders, rather than to a utility substation, are having a whole lot of trouble merely staying on-line."[5]

Harmonic Distortion

Another problem concerning the quality of electricity relates to *harmonics.* These are waveforms whose frequencies are multiples of the fundamental (60 Hz) waveform. The combination of harmonics in conjunction with the fundamental waveform produces a nonsinusoidal, periodic wave. Harmonics in electric utility transmission lines are a result of rapidly changing loads within the system.

There are voltage harmonics and current harmonics. *Voltage harmonics* can adversely affect customer loads, while *current harmonics* can adversely affect capacitor banks and telecommunications systems. If the current harmonics are of a sufficient magnitude, significant voltage distortion due to voltage drop can occur, especially in the absence of capacitors on the utility's system. If the current results in interference with communication lines (e.g., telephone companies sharing the same distribution system), then special filters may be required.

Harmonics are not a problem with synchronous and induction generators but they are with some inverters, particularly those of the line-commutated variety (discussed later in this chapter). The exact impact of harmonics is not fully understood. General harmonics standards have not yet been developed in the United States. However, total harmonic distortion (THD) values of 5 or 10 percent on current harmonics and 2 percent on voltage harmonics have been mentioned as preliminary maximum limits.

Excessive THD can be eliminated by adding filters to the power conditioning system. However, adding filters can be relatively complicated and expensive. It would probably cost less to buy a high-quality inverter with minimal THD than to purchase an inferior inverter and add harmonic filters to it.

Dedicated Transformers

In the distribution secondary, a number of customers may share a transformer. When this situation exists, it may be possible for an individual generator or a number of generators to carry the load in the distribution secondary and continue to operate even if this small section is cut off from the central power source via a switch at the transformer. This would not occur if the dispersed generator had a *dedicated transformer* connected in parallel with other distribution secondary sections along a distribution lateral. If the entire lateral is cut off from the utility, then the load applied to the generator is likely to be larger than its capacity. In this case the generator will be overloaded and its voltage will collapse (decrease), shutting off the generator. Many utilities require the installation of a dedicated transformer to prevent a dispersed generator from being isolated along with a relatively small load.

However, a dedicated transformer is expensive. Residential customers are normally served by a shared transformer paid for by the utility. If a dedicated transformer is installed with a residential cogeneration or small power production system, the customer is generally expected to pay for the transformer.

The characteristics of small, three-phase distribution transformers also complicate the dedicated transformer option. At present, three-phase distribution transformers under about 100 kw generally have a wiring arrangement which is ungrounded on the primary side. This makes it difficult to detect unbalanced conditions. Therefore, it may be necessary to use three single-phase transformers, instead of one three-phase transformer.

The dedicated transformer approach to preventing isolated operation with small induction generators and line-commutated inverters is possible only as long as there is not a high number of dispersed generators. If the dispersed generator capacity becomes large enough to carry the load along the distribution lateral, then protective relaying must be used to prevent isolated operation due to self-excitation.

Ground Faults

Ground-fault protection is much more complicated than protection against short circuits. Fortunately, separate ground-fault protection may not be necessary for small dispersed generators. Southern California Edison and San Diego Gas and Electric are among those utilities that do not require fault protection for units less than 100 kw. PG&E specifies ground fault protection only for those QFs of 40 kw and over.

These utilities believe that the existing ground-fault protection equipment at distribution substations is adequate for detecting ground faults from dispersed generators. This equipment will isolate the distribution feeder primary containing the faulted generator. Once the generator is isolated from the central power source, voltage and frequency relays on the generator will trip, and parallel operation under fault conditions is discontinued.

Ground-fault protection for three-phase generators is affected by the way the different phases are connected at the distribution transformer. With the wye or star connection, the center point is often connected to *ground*. (The ground is a reference point that remains at zero voltage.) A line connected to ground frequently extends to customers as well. If this is the case, two voltage levels result: the line voltage between any two nongrounded lines and the phase voltage between the ground line and a nongrounded line. This is how the common voltage combinations of 120/208 v or 240/415 v are obtained. With the delta connection, a fourth neutral line is not present, and thus there is only one voltage level.

For medium-size three-phase generators, the primary side of the distribution transformers is often delta connected. This makes ground-fault detection on the primary side difficult because there is no neutral line. (If a neutral line is available, ground faults are easily detected by monitoring the current level on the neutral line.) Without a neutral line, it is possible to detect faults on the primary side of the distribution transformer using either (1) a three-phase potential transformer with a broken delta-connected secondary and an overvoltage relay, or (2) a single-phase potential transformer with an under/over voltage relay. The latter method is likely to cost less but is not as reliable as the former.

As an alternative to measuring ground faults on the primary side of the distribution transformer, ground faults can be detected through measuring unbalance on the generator's side of the transformer. Unbalance detection involves using either a current or voltage balance relay. While the methods for ground-fault protection on the customer's side of the distribution transformer are less expensive than the methods employed on the primary side, the reliability of the detection on the customer's side is less certain.

Power Factor

Unlike the concerns above, low power factors do not harm equipment, but they can impose additional costs on the utility. When providing the consumer with electrical service, the utility is actually furnishing two types of alternating current (AC). The major type is known as the active or *real kilowatt component,* upon which the cost of service is based. This is the component that is usable and does the work. The other kind of energy is the reactive or *magnetizing kilovar component,* which is needed to magnetize any electrical equipment that requires a magnetic flux from the power system to permit operation. This type of energy does not cause the disk of the watt-hour meter to rotate, although it is drawn from the power lines and furnished to the system. Every piece of electrical equipment or appliance whose operation is dependent upon a magnetic circuit requires a supply of existing or magnetizing current. For example, a transformer or a common induction motor receives magnetizing current through the AC distribution systems.

The *power factor* can be defined as the ratio of the working current to the total of the working current plus the magnetizing current. A low power factor means that an excessive amount of magnetizing current is being drawn from the incoming power lines. During the past years when electrical power was cheap and in plentiful supply, the taking of magnetizing current from the AC lines was not considered objectionable if the effect of the extra current on voltage regulation was not too serious.

However, as the total electrical load approaches the capacity of the utility's generators, a low power factor will cause these generators to become overloaded, and additional expensive generation equipment may be needed. Moreover, a low power factor results in the overheating of conductors and transformers and low voltage throughout the distribution system. This results in the inefficient operation of electrical equipment and overheating of induction motors.

To clarify the admittedly murky power factor principle, consider the following example. The power flowing in a direct current (DC) circuit is the product of volts times amperes (amps) and is expressed in watts. A *wattmeter* similar to that measuring the flow of electricity into

your house will accurately register this value. In an AC circuit, if the
load is pure resistance, then again volts times amps equals watts.
However, if the AC circuit contains an induction motor, the wattmeter
reading is less than the product of volts and amps. A fraction of the total
amperage is consumed in magnetizing the motor, and the balance is
used to perform work. Consider the following example of a 240-v
motor drawing 10 amps as measured by an *ammeter.* The product of
volts times amps would equal 2,400 watts (w). However, when the
power is measured with a wattmeter, we find it registers only 1,920 w.
It is apparent the current is doing other than providing useful work. The
current necessary to provide the useful work is 1,920 ÷ 240 = 8 amps.
Since the total current is 10 amps, the power factor becomes 8 ÷ 10
= 0. 80 or as is usually stated, watts ÷ volts x amps = 1,920 ÷ 2,400
= 0.80 or 80 percent.

The crux of the matter is that the magnetizing current constitutes
an additional load on the utility without producing revenue. Many
utilities now compensate for this loss by including a charge for a low
power factor in their rate schedule. The power factor is also a function
of the phase angle between the voltage and current signals on a particular line. If voltage and current are in phase, then the power factor
equals one. If voltage and current are out of phase, the power factor is
given by the cosine of the phase angle. The power factor is said to be
lagging if the voltage is slightly behind the current and is said to be
leading if the voltage is slightly ahead of the current.

The various types of power are measured in different units. *Real
power* is given in kilowatts, *apparent power* in kilovolt-amperes (kva)
and *reactive power* in terms of the kilovolt-ampere reactive (kvar).

The size of synchronous generators (a fuller discussion of these
types of generators follows below) may be specified in kilowatts or
kilovolt-amps. The kilovolt-amp rating is the rated voltage multiplied
by the rated maximum current, and the kilowatt output is given by the
kilovolt-amp multiplied by the power factor. If a manufacturer rates a
machine in kilowatts, he must specify the power factor. A power factor
of 0.8 is normally used. Thus a 200-kva generator actually produces
160 kw, but could produce up to 200 kw if the power factor were
raised to 1.0.

Generator engineers talk about the flow of watts and the flow of
VARs (volt-ampere-reactive). The power output of a generator is controlled by varying the torque applied to its shaft by the prime mover.
The VARs output is controlled by varying the generator's field excitation. This function can be valuable in an industrial plant. If an in-plant
generator is overexcited, it produces VARs as well as watts, and these
VARs flow into the plant's motors to provide their excitation current.

This reduces the amount of VARs the motors draw from the utility system. It is equivalent to installing power factor correction capacitors (see below).

Certain components in the electrical system either generate or absorb reactive power. Generators of reactive power include synchronous generators, capacitors and lightly loaded lines. Absorbers of reactive power include transformers, induction motors and generators and heavily loaded lines. An electrical utility must balance the supply and demand for reactive power as well as the supply and demand for real power. Since induction motors represent up to 70 percent of the overall load in a utility system, there generally tends to be a shortage of reactive power.

Typically residential loads place a lagging VARs load on the utility's distribution system, which lowers the power factor. This reactive power produces no revenue for the utility but does produce real power losses in its system. These needed VARs consumed by the QF are usually supplied by the capacitor bank nearest the lagging VAR load.

Residential power factors are relatively unimportant. Residential customers are billed for true power consumed. The rates they pay are already adjusted for the loss of power the utility incurs. The need for calculating or measuring kvar and kvar hours (kvarh) is thus avoided after the initial determination of residential power factor is made. Residential power factors fall in a range between 0.80 and 0.95. When compared to unity (1.0) power factor, a 0.85 power factor increases kilovolt-amp loading approximately 15 percent and adds a VAR load equal to about one-half of the kilowatt load.

Special metering is sometimes needed to determine the power factor of certain industrial processes. The utility may either charge a penalty for a low power factor or refuse service until the industry takes measures to adjust the power factor to meet the utility's minimum standards. Utilities normally require dispersed generators to meet minimum power factors (e.g., 85 percent).

Generators, Inverters, Self-Excitation and Power Factors

AC generators and inverters are devices that produce electrical energy in a form compatible with utility systems. Generators are rotating machines driven by engines or turbines, while inverters are electronic devices that convert direct current into alternating current. Inverters are often used in wind, hydro and photovoltaic power systems.

The *synchronous generator* is the machine used to generate the vast majority of AC power today. It contains an exciter, a DC field winding, armature winding and mechanical structure. In all but the smallest synchronous generators, the field winding is part of the rotor, and the armature winding remains stationary.

The rotor, or rotating armature, is driven by a prime mover (either a turbine or engine), and as the magnetic flux generated by the DC field windings crosses the stator windings, a three-phase voltage and current is induced at the output terminals. The exact value of the AC voltage generated is controlled by varying the current in the DC field windings, while the frequency is controlled by the speed of rotation. Power output is controlled by the torque applied to the generator shaft by the prime mover. Thus, the synchronous machine provides very precise control over the power it generates.

Nearly all synchronous generators are three phase with voltage outputs ranging from 120 v to 13,800 v. The full load efficiency of synchronous generators is typically 83 to 96 percent for medium and high speed (less than 200 revolutions per minute) machines in the range of 100 kw to 10,000 kw. The efficiency of synchronous generators increases with generator size and declines as the load decreases. The percent loss of efficiency from the full load level is typically 2 percent at 80 percent of the rated load, 4 percent at 60 percent of full load, 7 percent at 40 percent of full load and 11 percent at 20 percent of full load. The power factor for synchronous generators varies from about 0.80 lagging to 0.90 leading.

Since synchronous generators have an exciter, they can operate without an outside source of power. However, the capability for independent operation necessitates regulating the voltage and frequency outputs from the generator. Voltage and frequency control are accomplished using a voltage regulator and speed governor. Also, when two or more synchronous generators are connected in parallel, their phases must be synchronized. Thus, synchronous generators connected in parallel to a utility system must include a synchronizing device.

The cost of synchronous generators decreases from about $100 per kilowatt for generators of about 10 kw to about $20 per kilowatt for those of more than 1,000 kw. These costs do not include contractor installation and markup. The installation cost for synchronous generators is approximately equal to the capital cost.

The *induction generator* is a much simpler machine than a synchronous generator. It has very little control over its electrical ouput. It is exactly the same in design as an induction motor. It usually cannot generate electric power when isolated, but when coupled to the utility network, it draws a reactive excitation current from the line, and when driven, feeds AC power back into the line. Its advantages for small interconnection applications are that it cannot operate without excitation supplied by the utility system. Very little protective relaying is necessary, and speed control isn't needed at all. The excitation power supplied by the utility also modulates the voltage, frequency and phase of the signal produced by an induction generator.

Induction generator efficiency at full load varies from about 80 percent for small generators to about 95 percent for larger machines. The efficiency of induction generators, like synchronous generators, is relatively constant at part loads down to about 25 percent of full load.

The power factor for induction generators depends on the loading conditions. At full load, induction generators generally have a power factor of about 0.80 lagging. The power factor declines as the load fraction decreases. For example, a 340-kw induction generator has a power factor of about 0.70 at 50 percent load and 0.55 at 25 percent load.

In general, induction generators are simpler and have lower capital and maintenance costs than synchronous generators. For example, one generator manufacturer sells induction generators for about 20 percent less than synchronous generators. Due to their simplicity and lower cost, induction generators are advantageous at sizes on the order of 500 kw and less, if there is no need to operate without the utility. Above 500 kw, the increased power factor and efficiency can outweigh the added expense for an exciter and controls with a synchronous generator.

Many renewable-based technologies, such as wind and photovoltaics, generate direct current. This must be converted into alternating current if it is going to be used to run AC loads or be delivered to the grid system. This conversion is done with an inverter. They are available for generators of under 1 kw to 500 kw. The efficiency of inverters is generally in the range of 90 to 95 percent at full load. The efficiency generally remains above 80 percent at part loads down to 20 percent of full load.

There are two types of inverters: line-commutated and self-commutated (also known as forced commutated). *Commutation* is the switching process that produces the time varying signal. Line-commutated inverters require reactive power from the utility system or some other source in order to operate. Self-commutated or forced-commutated inverters, however, are capable of independent operation, as in a system with battery storage.

The performance of self-commutated inverters is generally superior to that of line-commutated inverters. While self-commutated

inverters can have a unity (power factor equals 1) or even slightly
leading power factor, line-commutated inverters often operate with a
relatively low, lagging power factor. One field test of an 8-kw, line-
commutated inverter in a photovoltaic power system yielded power
factors of only 0.36 to 0.72.

Self-commutated inverters generally produce THD levels of 5 percent or less, a level that is ordinarily acceptable to utilities. However,
excessive levels of THD may be generated with line-commutated inverters. For example, the photovoltaic system noted above operated
with a THD level in the current waveform of 28 percent.

At present, inverters are relatively expensive. Self-commutated
inverters designed for parallel operation with a utility typically cost
about $1,500 per kilowatt for a 5-kw application and $500 per kilowatt
for a 300-kw application. Line-commutated inverters can be significantly less expensive. A popular small commutated inverter (under 10
kw) costs $350 to $800 per kilowatt depending on the exact size.

Cheaper solid-state inverters generate square waveforms, which
can harm sensitive electrical appliances. These waveforms can be
smoothed out to sinusoidal shapes through the addition of filters.

Low power factors can be corrected, but some of the corrections
can themselves generate other problems. For example, *capacitors,* devices that temporarily store electricity, can be inserted into the system
to bring the voltage and current in phase. However, since the VARs
drawn by an induction generator vary from a no-load value to a peak
at maximum power output, they cannot be supplied entirely by static
capacitors. More important, it is possible for an induction generator
with capacitors to self-excite when the capacitor is connected and to
generate a voltage, even though it is disconnected from the network.
This nullifies the inherent safety feature of an induction generator, that
is, its inability to operate in isolation.

(continued on page 148)

Figures 4–1, 4–2, 4–3: The interconnection schematics shown on the following pages were developed by the Pennsylvania Power and Light Company to be used with systems of different sizes. The components shown
here are typical of what the Pennsylvania Power and Light Company has
determined to be necessary to protect both the utility and the small power
producer. These requirements represent almost a full menu of what your
local utility might demand, although many utilities require less protective
equipment. The small power producer can also play a role in determining
the interconnection requirements for the particular utility to which it will
connect. The standard numbers in these schematics are those used in *Electric Power System Device Function Numbers* (New York: American National Standards Institute, 1979). Order no. ANSI/IEE 37.2. *Redrawn from
schematics provided by Pennsylvania Power and Light Company.*

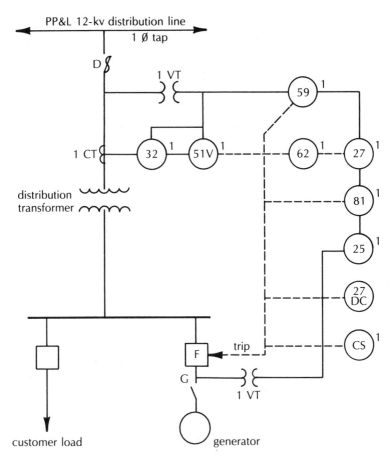

Figure 4–1
Interconnection Schematic: Single Phase up to 200 kva

D—fused disconnect
F—generator breaker or contactor
   (electrically operated)
G—safety switch (lockable in the
   open position)
CS—PP&L control switch
CT—current transformer
VT—voltage transformer

Intertie Protective Relay Types
25—synchronism check
27—undervoltage
27DC—battery undervoltage
32—directional power
51V—torque controlled
   time overcurrent
59—overvoltage
62—time delay
81—frequency

NOTE: Safety switch (G) must be located electrically between the generator
and the transformer. The specific location will vary based on
metering requirements.

Figure 4–2
Interconnection Schematic: Three Phase up to 1,500 kva

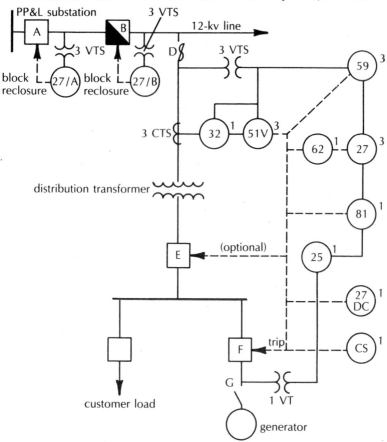

A—substation circuit breaker
B—electronic oil circuit recloser
D—fused disconnect
E—transformer low side breaker
(optional)
F—generator breaker
(electrically operated)
G—safety switch (lockable in
the open position)
CS—PP&L control switch
CTS—current transformers
VT—voltage transformer

VTS—voltage transformers

Intertie Protective Relay Types
25—synchronism check
27—undervoltage
27DC—battery undervoltage
32—directional power
51V—torque controlled
time overcurrent
59—overvoltage
62—time delay
81—frequency

NOTE: Safety switch (G) must be located electrically between the generator
and the transformer. The specific location will vary based on
metering requirements.

Figure 4–3
Interconnection Schematic: Three Phase up to 1,800 kva

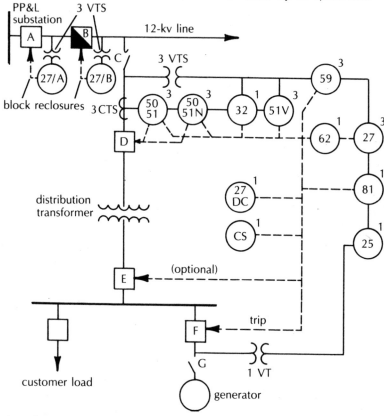

A—substation circuit breaker
B—electronic oil circuit recloser
C—disconnect
D—"point of contact" breaker
E—transformer low side
    breaker (optional)
F—generator breaker
    (electrically operated)
G—safety switch (lockable in
    the open position)
CS—PP&L control switch
CTS—current transformers
VT—voltage transformer
VTS—voltage transformers

Point of Contact Relays
50/51
50/51N

Intertie Protective Relay Types
25—synchronism check
27—undervoltage
27DC—battery undervoltage
32—directional power
51V—torque controlled
    time overcurrent
59—overvoltage
62—time delay
81—frequency

NOTE: Safety switch (G) must be located electrically between the generator
and the transformer.The specific location will vary based on
metering requirements.

147

The installation of capacitors is the standard method for increasing a lagging power factor. Capacitor banks are routinely installed throughout the distribution system for power factor improvement. Some industrial customers install their own capacitors near their facility. Large banks of capacitors on the distribution primary are much more economical than smaller units of capacitors distributed throughout the network near generators. If capacitors are installed centrally, the cost for the capacitor bank can be shared among those receiving power factor compensation. Also, as mentioned above, if capacitors are located on distribution secondaries near induction generators and line-commutated inverters, there is the potential for self-excitation.

For dispersed generators under 100 kw, a concern for power factor may be unnecessary if the small generators taken together have a negligible impact on the power factor of the whole utility system. Southern California Edison and San Diego Gas and Electric, for example, are not requiring dispersed generators of less than 100 kw to correct a lower power factor. However, if the penetration of small dispersed generators with low power factors becomes high, then it may be necessary to apply power factor requirements or penalties to small generators and inverters as well as to large industrial or commercial generators. If the power factor of a cogenerator or small power producer is of concern, it may be necessary to monitor power factor. Power factor can be determined by measuring reactive power or reactive energy. Reactive energy meters cost about $500 installed. Also, meters are available that measure power factor directly.

Interconnection Costs

Interconnection costs as a percentage of total project costs vary inversely with the size of the facility. The larger the power plant, the smaller the investment per kilowatt of installed capacity. Table 4–1 provides information on the total interconnection costs for the three sizes of cogeneration systems. The costs for hydroelectric, photovoltaic or wind-powered systems would not differ greatly. The costs for each system are divided into two categories: a best case, which assumes that much of the equipment is already in place or not required (e.g., capacitors, dedicated transformers, protective relays) and a worst case, which assumes this equipment must be purchased. Most of the cost difference comes from the addition of a dedicated utility transformer and more expensive protective relays and other protective devices.

Interconnection costs will be an important factor in deciding the economics of very small systems. Table 4–2 gives cost information for a 10-kva system under several interconnection schemes. Scheme 9.1 includes the dedicated distribution transformer, relays, capacitors at

TABLE 4–1
Interconnection Costs for Three Typical Systems

EQUIPMENT	50 KW BEST	50 KW WORST	500 KW BEST	500 KW WORST	5 MW AVERAGE
Capacitors for power factor	*	$1,000	*	$5,000	*
Voltage/frequency relays	$1,000	1,000	$1,000	1,000	$1,000
Dedicated transformer		3,900	*	12,500	40,000
Meter	80	1,000	80	1,000	1,000
Ground fault over-voltage relay	600	600	600	600	600
Manual disconnect switch	300	300	1,400	1,400	3,000
Circuit breakers	620	620	4,200	4,200	5,000
Automatic synchronizers	*	*	2,600	2,600	2,600
Equipment transformers	600	1,100	600	1,100	1,100
Other protective relays	*	3,500	*	3,500	3,500
Total costs ($)	$2,600	$13,020	$11,080	$32,900	$57,800
Total costs ($/kw)	52	260	22	66	12

NOTE: Reprinted from *Industrial and Commercial Cogeneration,* Office of Technology Assessment (Washington, D.C.: Office of Technology Assessment, 1983), p. 37.
*Indicates an optional piece of interconnection equipment that was not included in the requirements and cost calculations.

the generator and a time-of-day meter. This is the most expensive interconnection arrangement. If an ordinary energy meter is used rather than a time-of-day meter (scheme 9.2), the cost drops by 8 percent. Placing the capacitors on the distribution lateral rather than at the generator (scheme 9.3) reduces the total cost 15 percent from the first case. Compared to scheme 9.3, the cost savings is minimal if power factor correction is not included (scheme 9.4).

It may be possible to dispense with the dedicated distribution transformer for a small, single-phase generator or inverter. This would reduce the total interconnection cost by about 25 percent (schemes 9.5 to 9.7). Scheme 9.8, the least expensive arrangement considered, excludes the transformer and includes an ordinary energy meter.

TABLE 4–2

**Interconnection Equipment and Systems
for a 10-kva (single-phase) Induction Generator
or Line-Commutated Inverter**

INTERCONNECTION ITEM	INSTALLED COST (1981 $)	INTERCONNECTION SYSTEM SCHEME							
		9.1	9.2	9.3	9.4	9.5	9.6	9.7	9.8
Disconnect switch, 60 amp	176	X	X	X	X	X	X	X	X
Dedicated distribution transformer, 15 kva single phase	865	X	X	X	X				
Potential transformer, 240 v	241	X	X	X	X	X	X	X	X
Over/under voltage relay	436	X	X	X	X	X	X	X	X
Over/under frequency relay	531	X	X	X	X	X	X	X	X
Molded-case, thermal-magnetic circuit breaker, 60 amp	532	X	X	X	X	X	X	X	X
Capacitors, 4 kvar at the generator	621	X	X		X				
within the distribution system	41			X			X		
Meter ordinary kwh	80		X						X
time-of-day meter with dial registers	367	X		X	X	X	X	X	

Source: Howard S. Geller, *The Interconnection of Cogenerators and Small Power Producers to a Utility System* (Washington, D.C.: Self-Reliance, Inc., 1982).

Radial and Network Systems

Two different methods of distributing electricity characterize our electric system. Most of the country is served by so-called *radial systems*. In these there is only one line between the power plant and the customer. The radial system can be visualized as a giant trunk line that has branches and these in turn have subbranches.

The *network system* looks more like a grid system. Here a customer is served through many routes. Network systems are typical of high density urban areas. In many cities, about 15 percent of the service area is served by network systems. In New York City, more than 70 percent of Consolidated Edison's (Con Ed) service area is served in this manner.

Because of their redundant nature, network systems are more reliable than radial systems. However, because of their redundant nature they require more protective relays than a radial system might. Reverse power flows must be prevented from leaking into the system. For example, Con Ed has reverse relays that trip at power flows as small as 10 kw. As a result, dispersed generators in these areas must connect to the primary distribution line or forego selling electricity to Con Ed. These lines carry higher voltages and may be located far enough away from the generator that the interconnection costs rise greatly.

Other Utility System Performance Issues

If many dispersed generators come on-line, the utility might have to replace existing transformers, circuit breakers, cables or other components of the T&D system. That would be necessary if dispersed generators produced current or power at levels that exceed the capacity of existing equipment. Utilities are allowed to charge QFs for the cost of this upgrading.

However, a utility should not have to replace any equipment as long as the peak rating of the dispersed generators is less than the maximum on-site load. In addition, the impact that a dispersed generator has on distribution system capacity should be placed in the broader context of planned operations to the capacity of distribution system equipment.

Even if the generator rating is greater than the on-site load, the utility system will only receive the net power produced by the customer's facility. For example, if an on-site generator produces 100 kw, while other electrical equipment at the site of the generator consumes 75 kw, then 25 kw is the net amount of power delivered to the utility. It may be possible for the utility to use the maximum *net power output* for sizing distribution system equipment. Moreover, the installation of dispersed generators could effectively expand the capacity of existing utility equipment. Reliable QFs could allow the secondary system to serve more homes than normal. However, to do this, the reliability of the generator or load and the number of generators is critical. Utility equipment on the secondary or lateral could become overloaded when the dispersed generator is not operating if utility equipment has been sized based on the assumption that power from the dispersed generator is supplying part of the load.

Another issue related to utility system performance is the amount of fault current that dispersed generators can contribute at certain points in the distribution system. If this contribution is large enough, the utility will have to replace circuit breakers in order to upgrade the short-circuit capacity. One study of wind turbines at different penetration levels found that at the distribution substation, the wind turbines increased the short-circuit current potential less than 10 percent, even with a 50 percent wind turbine penetration. However, the potential short-circuit current at the ends of the feeder increased 10 to 60 percent with a dispersed generator penetration of 20 to 50 percent. Thus the ratings of circuit breakers and other overcurrent protection equipment near the end of the distribution system might be exceeded if dispersed generators reach a high level of aggregate or total capacity.

Another legitimate concern is that cogenerators and small power producers could increase the load factor on some equipment in the utility system. Highly fluctuating power sources such as wind generator and photovoltaic systems are likely to switch capacitors on and off more frequently than they would otherwise be. This would reduce the service life of the capacitors.

The issues and technologies related to interconnection standards are complex. Moreover, the technologies themselves are changing rapidly. The standards required will evolve and probably tighten as more and more dispersed generators come on-line. On the other hand, the technologies, especially those for protective relaying, will probably become cheaper, more reliable and more sophisticated as microprocessor technologies are introduced. This chapter has focused on the multiple issues related to interconnection. Chapter 5 focuses on negotiating with the utility. The question this chapter answers is, how can the QF get the best deal?

CHAPTER 5
Getting the Best Deal

The aim of the previous chapters was to provide a basic understanding of the economic and engineering principles underlying such issues as avoided costs and interconnection standards. This chapter deals only with practice and strategy. Avoiding the question of what should be, it tells what is and how qualifying facilities (QFs) can operate within the existing laws and regulations to negotiate the best contract.

Always remember, the utility is trying to get the best deal for itself. The negotiation process is to some extent an adversarial one, so be prepared. Don't sit down at the negotiations table without having done your homework, and don't negotiate by phone. Telephone conversations are fine for eliciting information, but serious bargaining can only be done face-to-face.

You must know your rights under the Public Utility Regulatory Policies Act (PURPA). You should also make sure the utility knows that the public service commission, the city council or the Federal Energy Regulatory Commission (FERC) is aware of your efforts. Do some background investigating to uncover the operating characteristics of the local utility. Find out how their PURPA tariffs were developed. Read through the existing contracts on file at the regulatory commission, the city council or the rural electric board. Know how your own technology works. Make sure you have performed a basic cash-flow calculation for your proposed investment so you know what the minimum acceptable terms are that allow you to go ahead with the project. And be prepared to bargain. Negotiations are a process of give and take.

Remember also that in the best of situations QFs are entering a field that involves risk. Negotiations may reduce the risk, but they cannot eliminate it. QFs must accept the risk that the technology may not function as expected. The wind turbine, for example, may be out of service more than expected. Or the wind regime may not be as strong as had been predicted. To that you must add the economic risk that avoided costs could drop in the future and reduce your revenue.

Finally there is the political and legal risk that court orders or legislation will dramatically change the QF's rights under PURPA.

The Regulatory Process

QFs have to bargain as individuals, but they also bargain as a class. This type of bargaining takes place in the regulatory commission chambers, city councils or state legislatures. As a previous chapter pointed out, the PURPA regulatory process is an ongoing one, and owners of QFs must participate in this evolution. To do that you should be well aware of the regulatory process for implementing PURPA. The FERC required state regulatory commissions and unregulated utilities to initiate PURPA proceedings by March 1981. Most of them missed the deadline, but they had initiated such proceedings by the end of 1981.

The first step was for commissions to appoint a hearing examiner or administrative law judge to take testimony and to hold hearings on possible rules to implement the federal law. This is not an evidentiary procedure, which means it doesn't involve witnesses presenting formal evidence. Nor does it permit intervenors to request data in support of witness testimony. The process mainly consists of comments filed by all parties. The hearing examiner then issues a decision. This process takes about six months (although it can take a great deal longer when there are intervening lawsuits, as in the case of PURPA).

The parties have a period to file exceptions and comments. The commission then issues a Proposed Decision. There is another comment period, and then the commission issues a Final Order Adopting Rules. This process usually takes about three months.

At the time the commission issues the Final Order Adopting Rules, it issues a Compliance Order to utilities under its jurisdiction to file tariffs in compliance with the Order. Within two to three months, utilities file draft tariffs. Some commissions routinely approve these. In many, these draft tariffs are vigorously contested and public hearings are held. These hearings, which can take up to a year, are evidentiary in nature. Utilities can be forced to justify their tariffs with data. Usually a hearing examiner or administrative law judge presides over these hearings. Sometimes the full commission does so. These hearings can take up to a year. Another comment period follows and then the commission issues a final order approving tariffs with certain modifications.

Thus, even in the best of circumstances, the entire process takes about 18 months. The New Hampshire Commission combined the rulemaking, ratemaking and evidentiary hearings. Spurred on by the passage of the state's own mini-PURPA law in 1979, the commission completed its task within 6 months. No other state has been as efficient.

In some states, actions by the legislature have forced commissions to start all over again. Indiana and Minnesota have reopened the implementation process. In other states, like Arkansas, the administrative law judge issued an opinion on proposed rules in August 1982. Rhode Island is reopening its process. In California so much controversy has surrounded the PURPA rules that it is in its third year of hearings. Utilities are still operating under interim tariffs. "It's driving me nuts," says one wind developer.

Many QFs are frustrated at the apparent slowness with which PURPA is being implemented. Yet the rule has been that the more rapid the implementation, the worse the buyback rates and other terms QFs receive. (New Hampshire and Vermont are exceptions to this rule.) States in which commissions routinely approve draft tariffs submitted by utilities, or that hold no hearings because there are no QF intervenors, usually provide for very low buyback rates and discriminatory standby or back-up charges. Those states that are just entering their PURPA implementation process can learn from other states. QFs who in 1979 had little knowledge of utility economics and the concept of marginal cost pricing are now much better versed in these concepts.

Sometimes utilities insert provisions in contracts that allow them to terminate the contract if pending legal suits against PURPA are successful. Thus, for example, in the preface to a contract the utility may insert the following:

> *WHEREAS, Buyer is entering into this Agreement in good faith in compliance with article 210 of the Public Utility Regulatory Policies Act of 1978 (16 U.S.C.A. 824a-3), and (applicable state statutes) and regulations promulgated thereunder, in order to lessen Buyer's dependence upon foreign supplies of fuel oil and the uncertainties inherent in such foreign supplies;* without waiving any claim it may have as to the validity of said laws and regulations.[1] *(Emphasis added by author.)*

This bail-out provision is not normally found in power contracts that are negotiated without the specific imposition of the PURPA avoided cost rate. The utility may also insert in its termination section clauses such as the following:

> *If after this Agreement becomes effective Section 210 of the Public Utility Regulatory Policies Act (or the relevant state statute implementing PURPA) is repealed or modified so that Buyer is not re-*

quired to purchase energy and capacity at avoided cost, Buyer reserves the right to terminate this agreement.[2]

The seller should very much oppose these types of clauses. Instead, the seller should try to insert an antibail-out or positive provision that says, in effect, that the agreement remains in full force and effect even if PURPA is repealed or modified.

The possible insertions of bail-out provisions make it extremely important that QFs form associations, as QFs in a dozen states have already done, to lobby the legislature and to provide expert testimony to the public service commission.

One possible outcome is a state mini-PURPA law that mirrors the federal law. In that way the state's small power producers will be largely immune from court reversals of federal regulations. Another is the direct intervention of the state legislature in establishing rates or other contract decisions. This is rare but not unprecedented. At least two state legislatures have been so frustrated by the regulatory commissions that they established high minimum buyback rates *before* the commissions issued their own final tariffs. Statewide associations of QFs facilitate the sharing of information about existing contracts, which puts QFs in a better position to negotiate the best deal. This sharing of information can also expose patterns of discrimination on the part of utilities, which could prompt regulatory commissions to hold evidentiary hearings in which QFs as intervenors in the hearing can ask for specific data from utilities.

Many QFs have found it easiest to negotiate with utilities that have independent oversight agencies. The latter can be used to cajole or coerce the utility into offering better terms. Most state regulatory bodies lack oversight authority over municipal utilities and rural electric co-operatives (RECs). But the Oregon and Minnesota legislatures expanded the jurisdiction of the public service commission to include co-ops, public utility districts and municipal utilities for the purpose of implementing PURPA Sections 201 and 210.

Where no independent oversight function exists, one must depend on the city council or the rural electric board to play that role. City councils lack the expertise and, depending on the city, may be hesitant to become involved in QF negotiations or tariff development. On the other hand, they are vulnerable to direct political pressure in a way that regulatory commissions are not. Rural electric boards must play the dual role of manager and customer. Beset by their own problems with cost overruns and changes in demand, many view QFs as an additional nuisance. On the other hand, many of their rural customers find hydroelectric and wind facilities economically attractive. These customers are also owners, and as they also become producers a political constituency can grow within the association to make its presence felt. One farmer who owns a wind turbine told about his problems with the rural electric co-op, but added, "There are another ten machines going in real soon. When that happens I expect the co-op will be a lot more generous in its dealings."

In any event, the owner of a QF should understand that his or her individual bargaining session is set against the backdrop of vast changes in the electric system. This country is in the process of deregulating and decentralizing the electric generation system. Rate structures are being revised, load profiles are changing and new technologies are coming on-line. The QF owner must learn to live with the risks involved in such a fluid situation.

The Negotiation Process

For those with sufficient courage to push on, the following story should provide a sobering example of what could happen. Consider the trials and tribulations of one pioneer, Robert Hetzler, owner of a 20-kilowatt (kw) wind system in Mandan, North Dakota. His Rural Electric Cooperative (REC) demanded that he get certified by the FERC even after he showed them the regulation that said mandatory certification is only necessary for QFs over 500 kw. Then they called a meeting with him and the state electrical safety board even though the safety board has no jurisdiction over co-ops. "At the meeting the co-op literally demanded that the electric board stop the project on safety grounds," he remembers. The electric board refused. The co-op then demanded the board send an inspector to certify the wiring. They did and the inspector found the wiring "perfectly acceptable." The co-op pays Hetzler 1.3¢ per kilowatt-hour for any power he sells and charges him 5.2¢ per kilowatt-hour for any power he buys. They billed him $700 for the hookup, including mileage for the trip out by the co-op's engineer. Then the REC insisted he carry $1 million in general liability insurance plus $500,000 in property insurance. "They came by and pulled out my meter when they decided I didn't have it [insurance]." Then they claimed his meter was defective because he couldn't be generating so much electricity, and they replaced it twice. The meter tested out perfectly. Then the REC refused to pay him $42 they owed because they first wanted a letter spelling out Hetzler's "payment policy." A beleaguered but still fighting Hetzler rhetorically asks, "Now that's discouragement, isn't it?"

Hetzler's experience isn't unique, but he's a pioneer. This book is based on the experiences of the first 200 small power facilities to come on-line under PURPA. These people have cut a path through the jungle. Those who come later will have their route marked and an easier path to follow. And those who come still later may find a wide highway with all of the bumps flattened out. In fact, even the pioneers note changes in the attitudes of utility personnel over the periods of their contracts. Initial hostility has given way to mutual respect and even warmth. A number of the pioneers express satisfaction and even gratitude for the help they received from the utility. For after all, utility people are also breaking new ground.

For almost all QFs, the negotiation process is a learning experience. For those willing to take the time to do some basic research, a little homework can make an astonishing difference in the outcome of bargaining.

Take the case of John Eckland of Great Falls, Virginia, and his local utility, Virginia Electric Power Company (VEPCO). In May 1980 VEPCO offered independent power producers a purchase price of less than a penny per kilowatt-hour. It levied an astonishing monthly charge ranging from $40 to $90 for those interconnected with its system and insisted that residential customers choosing to generate electricity switch to a highly discriminatory rate structure. Eckland tried to discuss some changes in this rate structure with VEPCO. When he received no response, he examined the FERC's regulations and then wrote to VEPCO's president, sending copies to the state utility commission and the attorney general of Virginia. The letter identified several apparent violations of federal regulations by VEPCO.

Not long afterward, VEPCO's rate department agreed to negotiate. After several meetings, Eckland won a fivefold increase in his purchase price and a tenfold reduction in the monthly service charge. He also persuaded VEPCO to eliminate its discriminatory rate schedule for independent producers. The moral of the story? "Take your time, read the regulations and don't be afraid of direct negotiation," advises Eckland. "It can work wonders."

If direct negotiations prove unworkable, the consumer can appeal directly to the state utility regulatory commission and, finally, to the FERC. The applicant may find a sympathetic ear. Court proceedings are also possible in the event of stalled negotiations. They are also available if regulatory or other commissions fail to enforce, or erroneously enforce, PURPA regulations. Most regulatory commissions are closely monitoring the way utilities respond to PURPA. When Pacific Power and Light (PP&L) appealed an order of the Montana Public Service Commission because it made complaint procedures available to QFs but not to utilities, the commission responded, "As the stronger negotiating party, the Commission expects that utilities can, and will, protect their interests. If in doing so, the utility requires terms considered onerous by the QF, or refuses to contract with the QF, the natural course of events would be a complaint to the Commission from the QF. . . . Complaint procedures contemplate actions from one aggrieved. In the context of interconnection of QFs, it is difficult to contemplate a time when a utility would be aggrieved in the negotiating process."[3]

QFs who believe the contract offered by the utility violates the final order of the state commission should complain before signing the contract. It is much more difficult for regulatory commissions to revise existing contracts than to intervene before the contract is signed.

Regulatory commissions have weapons beyond a complaint procedure. Several commissions have warned utilities that continued foot-dragging could affect future rate requests. The Montana commission reminded utilities "of their obligation to provide information to the Commission regarding their initial written response to each prospective QF. . . . In addition to reporting each *contact* made, the Commission directs the Companies to submit one copy of the completed contractual agreements with each QF. The Commission welcomes additional information that will aid the Commission in analyzing the individual efforts of each utility in encouraging QF contributions to a utility's resource base."[4]

When the California Public Utility Commission decided Pacific Gas and Electric (PG&E) was not encouraging cogeneration, it levied a $7.2 million penalty in a rate case. That amount was withheld contingent on PG&E's bringing on-line specific amounts of cogeneration capacity within a certain time frame. Montana's commission warned that "utility failure to actively pursue QF contributions to their resource base . . . (will) constitute failure to provide cost effective service. To the extent the evidence in future proceedings does not demonstrate that the utilities have in fact vigorously pursued such contracts, the Commission will use such evidence in considering whether utilities are providing adequate service at just and reasonable rates."[5] Thus, if sufficiently angered, a commission can impose severe financial penalties on utilities that do not negotiate in good faith.

Know Thy Opposition

Before entering negotiations, the QF should obtain as much information as possible about the utility with which it is bargaining. PURPA Section 210 requires that utilities provide specific data with respect to avoided costs. They must provide their avoided energy costs at the time of filing and projected costs for the next five years; they must provide information on their plans for additions to and retirement from generation equipment and purchases of firm power for the next ten years and they must estimate costs of capacity to be added to the system as well as the energy costs associated with operating new generation capacity. Regulatory commissions can add to this basic list. Montana, for example, requires its utilities to provide information on estimated line losses.

By 31 May 1982 and each two years thereafter, utilities with annual sales over 500 million kilowatt-hours (kwh) must provide such data. Utilities with annual sales of under 500 million kwh need to file similar data only upon request from a QF, and utilities receiving power through an *all-requirements contract* may file data on their supplier that is adequate to determine avoided cost. Most municipal utilities and RECs are of this latter type.

Supplementary data can come from several sources. One is the annual report that each utility is required to file with the FERC. Form 1 was discussed in a previous chapter in an example about Boston Edison. Form 12 also contains considerable amounts of data on the utility's cost structure. These reports are on file at the state Public Utility Commission (PUC). Find out from PUC staff whether the utility is currently involved in a rate case or if it is seeking a Certificate of Public Convenience and Necessity (CPCN). The latter is needed any time power plants are constructed or major changes are made in physical facilities. In either event, utilities will have to have filed considerable amounts of data to support their case in a hearing. This could be useful in evaluating avoided costs.

Another part of PURPA, Section 133, requires larger utilities to file cost-of-service data. This section was created to encourage regulatory commissions to revise rate structures to reflect marginal cost pricing, such as time-of-day rates and inverted rate structures. Note that the greater the success of Section 133 in meeting its objective, the less the difference between average (or embedded) costs and marginal or avoided costs would be. Utilities were required to submit detailed cost-of-service data for each customer class. These "133 Filings" should be compared to the avoided cost data which are filed separately under PURPA Section 210. These filings will include many load curves that can help QFs analyze peak and off-peak demand for different customer classes.

Examining Existing Contracts

Although one might expect'that contracts negotiated by the utility and QFs would be available for public inspection, they usually are not. The policy varies greatly. New Jersey, for example, will provide all documentation, including contracts and testimony at utility hearings, free of charge as part of their overall policy of strongly encouraging cogeneration. Most, like California, consider the contracts private. New Hampshire does also, but will divulge the names of the QFs who have signed contracts so they can be directly contacted. North Carolina and New York consider QF contracts public.

Commissions almost always charge for copying contracts or any other documentation. The usual charge is 10¢ to 15¢ per page, but this varies greatly. Virginia's commission charges 50¢ per page, and Florida's charges 75¢ to $1 per page. Given that one may need hundreds of pages of information to get a good idea of the utility's avoided costs, operating structure and other PURPA-related information, that can obviously put a large dent in the QF owner's pocketbook. You can avoid excessive copying charges by going to the regulatory commission office. Plan on spending a day or two at the regulatory commission in any case, talking to staff, using their library and examining documents.

PURPA requires all utilities to offer standard contracts to QFs with less than 100-kw capacity. This recognizes the unequal bargaining power of an owner of a 10-kw wind turbine against that of a multibillion dollar utility. The 100-kw cutoff is arbitrary. Some states have significantly raised it, which is specifically permitted by PURPA regulations. The presumption by FERC was that at $1,000 per kilowatt, the developer above that level would have sufficient money at risk that hiring expert assistance would not constitute a heavy additional burden.

The standard contract is developed by the utilities in close consultation with the regulatory commissions. Even though this contract is supposed to integrate all the attractive features of PURPA, QFs should examine it closely before signing. Some utilities have added burdensome conditions to their standard contract to encourage QF owners to negotiate one that removes these terms, while offering a lower buyback rate. After reading sample standard and negotiated contracts sent to QF owners by Southern California Edison (SCE) in 1982, the staff of the California PUC concluded, "Edison has aggressively encouraged private developers to accept a negotiated contract rather than a 'standard offer' at full avoided cost." SCE's standard contract, but not its negotiated contract, contained the following conditions: (1) Seller will reimburse Edison within 30 days for a portion of the capacity payments he has received if the project terminates before the end of the contract, (2) Edison has the right to take over the project and sell the electricity to itself at a low average rate if the original project operator abandons it and (3) Edison can renegotiate the terms of the contract at any time.

David Silverstone, a Connecticut attorney who has negotiated contracts on behalf of several QF owners, proposes some rules of thumb for prospective developers. "If you have a capacity under 200 kw, take the posted tariff," he advises, unless you absolutely need a long-term contract for financing. The long-term rate will be lower, and "you'll have to give up something to get the certainty of payment. Expect to take two to three months minimum for a signed contract for any load. Four to five months for any negotiated rate is very optimistic. It takes on average two to three months to get a letter of understanding."

As for legal services, Silverstone says that some clients want a lawyer only to look over and advise on a contract. Others want the lawyer to sit in on negotiations and cut the deal, using his or her knowledge of the utility business. The latter approach costs a great deal. Depending on the size of the project and complexity of the deal, it could take 50 to 100 hours of a lawyer's time, which could cost $75 or more an hour. Some lawyers will do the legal work on an "if come" basis, requiring payment out of the proceeds of electric sales on a monthly basis. Getting a hydro permit (discussed in a later chapter) can sometimes be done by clients on their own, without substantial attorney's time. Problems develop when there is a fight for site development rights. For 200-kw-and-under projects, "once there's a legal struggle over rights, the project loses its economic viability," says Silverstone.

Some suppliers of power plants will negotiate on behalf of their customers. Ken Hach is a dealer in Elkader, Iowa, of 10-kw Jacobs wind-power systems. He sold 10 in Iowa, 4 in Illinois and 4 in Nebraska. Sixteen of the 18 are interconnected. As part of the customer service, he handles all utility relations. His standard procedure is to notify the utility after the sale. Otherwise, utilities have been known "to try to discourage potential customers." Then he sends a letter to the chief engineer of the Iowa Commerce Commission, if the QF is in that state, and one to the FERC, notifying them of self-certification.

Standard Contract Variations

Everything in a contract is negotiable. If the QF agrees, even the provisions of the standard contract can be changed. However, once one provision is altered, the utility has the right to reconsider all other provisions. By altering one, the QF enters into an individual negotiating stance under the PURPA regulations.

QFs can choose to sell electricity on a short-term basis. They then sell it when available and receive a price based on the cost of displacing energy from the most expensive power plant operating at that time. In some states, like New Hampshire, the price the QF receives is the same no matter when it sells the electricity. In the vast majority of states, however, the price paid for electricity generated during the peak hours of the day and year is greater than that generated during off-peak hours. QFs who do not choose to install time-of-day meters are paid an average, which is either midway between the peak and off-peak rate or just the off-peak rate. If they want to be paid based on the time they generate electricity, they must sign a firm energy contract and install several registers, depending on the number of peaking periods. In California, which has three periods, peak, middle and off-peak, three registers must be installed. Each additional register costs $50 above the $50 regular cost of a watt-hour meter.

Often utilities will charge QFs for additional registers or meters. A number of commissions, such as Colorado's, have explicitly required utilities to charge the customer no more than its own cost. PURPA requires the charge to represent only the *additional* costs incurred by the utility. Thus it should be charging the customer based on its bulk purchase price, not the individual retail rate. Customers should also be permitted to purchase their own meters. In at least one case in Iowa, a customer convinced the commission to reduce the proposed metering charge simply by showing them a catalog with a very low price for such a meter. Some utilities have charged $15 to $80 a month for meters. These charges would allow it to recoup its costs within the first year or less. QFs should appeal such charges to the regulatory commission.

QFs can sign long-term contracts for energy and capacity. In those states where utilities have convinced the regulatory commission that they have an excess capacity and no construction plans for the foreseeable future, the QF is ineligible for capacity credits.

QFs willing to sign long-term contracts can usually gain a higher price for energy. In states that include a capital cost in their long-term energy cost calculations, as do North Carolina and Montana, the difference in price can be quite substantial.

Where capacity credits are given, they vary from $25 to $85 per year per kilowatt. In part, the difference is accounted for by the difference between the short-term capacity displaced, represented by the cost of an efficient combustion turbine that costs between $350 and $650 per kilowatt, and long-term capacity credits, represented by the cost of a baseload plant that costs between $1,200 and $3,000 per kilowatt. Part of the difference stems from the various methodologies used by regulatory commissions, including the various discount rates and inflation rates assumed.

About a half-dozen states give QF owners capacity payments in the form of a cents-per-kilowatt-hour rate for electricity delivered during peak periods of the year. In some states, like North Carolina, QF owners are only eligible for these "as available" rates if they sign long-term contracts. In Montana the minimum term is four years. In North Carolina it is five years. In others, like California, the QF can receive as available capacity payments even without a long-term contract. Such payments are based on the avoided cost of a combustion turbine. If the QF should want to receive the higher avoided cost based on a baseload plant, it could sign a long-term contract with the utility. Such a contract, however, would require the QF to meet certain performance criteria.

Long-term contracts usually trade off price for certainty. The QF might, for example, negotiate a fixed rate for energy and capacity credits. Since these are fixed, the QF knows what it will receive in the future. But since they are fixed, their real value decreases as inflation eats up future dollars. To alleviate this problem, the QF might negotiate a contract that has a fixed rate with a minimum floor below which the payments cannot drop or an escalator clause based on the consumer price index or the price of oil. A utility would negotiate such a contract only if it received something of value in return.

Many QF owners who are involved with outside investors or bank financing prefer to receive higher payments in the early years and lower payments in later years. Such contracts are usually called front-loaded contracts. The contract between U.S. Windpower (USW) and PG&E in 1982 illustrates such a contract.[6] A Massachusetts-based corporation, USW manufactures wind turbines and develops so-called wind farms consisting of dozens or even hundreds of wind turbines spread over a specific geographic area. The wind farms are excellent tax shelters because investors can buy into a specific wind turbine just the way they can invest in a specific oil or gas well and take advantage of attractive tax benefits associated with such investments.

USW needed a long-term contract to protect its investors. PG&E was willing to provide such a contract if it received benefits during the later years of the contract for its ratepayers.

During the early years of the 30-year contract, PG&E pays USW a constant price for each kilowatt-hour of electricity actually delivered. The price is 9¢ if the current Internal Revenue Service (IRS) provisions regarding tax credits continue in effect and 10¢ if the 10 percent investment tax credit for wind machines is eliminated. The constant price gives USW investors a guaranteed minimum (assuming the wind turbines themselves function as projected and the wind regimes are as estimated) and will probably be higher than the standard rate in the early years. PG&E keeps track of the difference between the constant price and 97 percent of the standard offer price in a separate account called a *payment tracking account* (PTA). The balance in the PTA account accrues interest at 120 percent of Bank of America's prime rate for 90-day loans to commercial borrowers. As the standard rate rises above the constant rate, the balance falls. PG&E estimates that the balance will fall to zero within 5 years. After the balance falls to zero PG&E pays USW only 95 percent of the standard offer price for energy delivered. This continues until 1 January 2002. Afterward, a 10 percent discount on the standard offer rate is in effect until the contract expires on 31 December 2011.

The contract contains no provision for capacity payments, because wind machines are not yet given any capacity credits in California. If these are added by the commission at some future time they will be included in the contract.

This contract provides USW a higher initial price for its electricity and a certainty about the price it will receive. That satisfies its investors. PG&E estimates that over the life of the agreement the net present value of ratepayer savings will be $5.8 million, representing an internal rate of return of 63 percent. In other words, the investors in the wind farm are willing to forego almost $6 million in revenue in return for the certainty of the long-term pricing arrangement and the higher prices in earlier years.

Long-term contracts benefit renewable-based technologies, like wind and solar and hydro, differently than they do technologies that use fossil fuels, like cogeneration. Renewable-based projects have high initial capital costs and low operating costs. Fuel is free. Given outside financing, the project developer not only knows the carrying cost but is paying a fixed monthly payment over 5 to 25 years. The fixed contract guarantees the QF and the bank that finances the facility a monthly payment that will meet the financing costs. (It is important to note here that no bank so far has lent a QF owner money when the only collateral was a long-term contract. The bank will require a net worth analysis of the individual's wealth and possibly other collateral as well.) As noted above, long-term contracts at fixed prices without a built-in inflation escalator clause decline in real value in later years. A dollar earned 5 years from now is likely to be worth far less than a dollar earned this year. But the developer may be willing to accept this decline in real values because the fixed monthly loan payments are also declining in real value. The loan is also being paid back in devalued dollars.

The cogenerator, on the other hand, has a low capital cost but a high operating cost. Fuel represents the primary component of that operating cost. Given uncertain oil prices and the deregulation of natural gas, the cogenerator would be unlikely to benefit from or to seek a long-term contract (unless methane were the fuel source or unless it were a bottoming-cycle cogeneration system operated by waste heat from an industrial process).

If a cogenerator signs a long-term contract to attract bank financing, it probably will still need to get insurance from another source to satisfy the bank. David Silverstone tells of one negotiation that illustrates this point. His client owned an 8-megawatt (Mw), coal-fired cogenerator sited at an abandoned mill yard. The developer is converting the mill to an industrial park and wants to sell steam to the tenants. Negotiations began in February 1981. The letter of intent from the local utility was finally sent in July for an avoided cost price of 6¢ per kilowatt-hour with a consumer price index escalator with an upper limit of the then-current price of oil in any year. The utility insisted on renegotiating the contract in the fall of 1981, partly due to the declining oil prices. It then announced it would only commit to purchase 75 percent of the output. So after a full year of negotiations, the cogenerator abandoned negotiations with this utility.

A contract was signed later with another utility. This utility refused to take the risk of rapidly rising oil prices increasing avoided costs, so the 15-year contract is tied to a percentage of avoided costs which varies inversely with the price of oil. The formula for estimating the percentage is the following: 1 divided by the square root of the current price of oil (O_c) minus the base price of oil (O_b) at the start of operation ($1 \div \sqrt{O_c - O_b}$). Thus, if the price of oil rises \$2, the formula equals 1 divided by the square root of 2 ($O_c - O_b = 2$), or 1 divided by 1.4, or 0.7. Thus the QF receives 70 percent of the current avoided cost level. To protect against oil prices dropping, the bank required the QF to take out a policy with an insurance company.

Remember the time value of money when negotiating. Know the difference between *nominal* and *constant* dollars. For example, a contract may agree to pay the QF owner the present value of future avoided costs. Assume the buyback rate today is 5¢ per kilowatt-hour and it rises to 7¢ per kilowatt-hour by 1985. Assume further that the consumer price index also rises, by 10 percent a year. Over three years the buyback rate has risen 40 percent, but inflation has eaten away almost all of that increase. So in constant dollars the QF hasn't made any more money.

PURPA does not require a QF to sign a contract based on its own production costs. However, QFs can negotiate such cost-plus contracts if they wish. Such a contract has been negotiated between the University of Washington and Seattle City Light and Power, a municipally owned utility. The agreement provides for separate purchase prices depending on the type of fuel the university burns in its cogeneration plant. When coal is used to fire the boiler, City Light will, according to the contract, purchase the electricity "at a price of 32.58 mills per kilowatt-hour, based on the University's estimated cost to produce the energy as determined in the attached Production Costs of University Cogenerated Power, Plant Operational Mode I, 14 October 1981." The price increases to 42.10 mills (4.21¢) per kilowatt-hour when oil or gas is the fuel. The purchase prices are based on the production costs plus an agreed-upon rate of return.

The first option is the standard rate, where over the life of the contract the QF is paid a standard rate that varies from year to year. Assuming no real inflation in the cost components used to compute standard rates, curve B in figure 5–1 would represent this option.

The second option is where the QF is paid a fixed minimum, in this case the greater of the initial year's standard rate (curve A) and the actual standard rate. In this case the rate can never drop (in current dollars) but could rise.

The third option is where the QF receives a fixed payment consisting of the initial year standard rate plus the previous year's inflation. Curve B represents this option.

The fourth option also consists of a fixed payment. In this case the QF is paid the initial standard rate plus a negotiated level of projected inflation. Curve C represents this rate assuming the projected inflation rate were to be higher after 1995 than the actual inflation rate.

The fifth option is a levelized payment that takes into account projected inflation rates. The present value of projected inflation (curve C minus curve A) is levelized over the life of the contract and added to the initial standard rate, resulting in curve D.

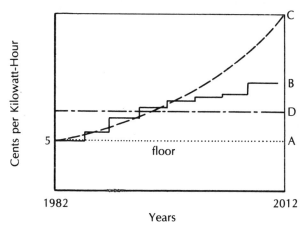

Figure 5–1: *Redrawn from* Public Service Commission of the State of Montana, Docket No. 81.2.15, *order no. 4865b, p. 16.*

Initially Montana's Public Service Commission merely recommended these as possible options. Given the reluctance of Montana utilities actually to offer these options, it held hearings in early 1983 to decide whether to make it mandatory for utilities to offer them.

Wheeling

While the QF owner cannot require the local utility to transmit electricity for sale to another utility, it might be able to make the transaction attractive enough to entice such a commitment. This is especially the case for those with municipalities or all-requirements utilities. Study the demand patterns, construction plans and avoided cost schedules of surrounding utilities.

In the case of the 8-Mw, coal-fired cogenerator previously cited, this QF ended up selling its electricity to a remote utility. The utility charged $25 per kilowatt per year for wheeling. For the 8-Mw facility, this amounted to $200,000, an attractive profit for the local utility. A 200-kw QF would be paying $5,000. The QF owner can afford to pay this amount if the price received from the next utility offsets that additional cost. Assuming a capacity factor of 80 percent for the cogeneration facility, the additional revenue comes to more than $500,000 a year for every penny increase in the buyback rate. A 200-kw QF with the same capacity factor would receive about $14,000 in additional revenue.

Some utilities have adopted a policy toward wheeling. Concord Electric Company, Exeter and Hampton Electric Company, and the New Hampshire Electric Cooperative, for example, have stated their willingness to wheel power from small generators for sale to Public Service Company of New Hampshire. In such cases, Public Service Company pays the small generator directly for the wheeled power.

The last chapter discussed the situation in Oregon with respect to wheeling. To provide higher buyback rates to QFs, Oregon has originated the idea of a base-case price. Even if a utility buys cheap electricity directly from the Bonneville Power Administration (BPA), it cannot use that price as its avoided cost. It must pay the QF either its avoided cost or the base price as calculated by the Public Service Commissioner, *whichever* is higher. The base price is high enough that it acts as an incentive for the local utility to wheel electricity to its supplier, thus giving the QF a higher price.

There is a further wrinkle to the wheeling issue in the New Hampshire Limited Electrical Energy Production Action of 1979. PURPA doesn't require utilities to wheel a QF's power to another customer, only to another utility. But the New Hampshire law allows the QF to demand that its utility wheel to as many as three customers so it can sell to them retail. As of late 1982 the law had never been tested. Many believe its provisions, which require rather than permit utilities to wheel and allow QFs to sell electricity directly to retail customers, are illegal.

Net Sales versus Simultaneous Purchase and Sale

The QF can choose one of two contractual arrangements. Under a *net purchase and sales contract,* the QF uses its power on-site and sells the excess to the utility. It buys power only when it cannot meet its own load. Sometimes in this case utilities will allow the QF to use an existing watt-hour meter. The current is reversed if the QF is selling power to the utility. Other utilities worry that reversing the meter will shorten its life and install a separate meter to measure the QF sales. In either case it is a net billing arrangement.

The QF can also choose to sell all of its generated power to the utility and simultaneously purchase all the plant's power needs from the utility. This arrangement is generally known as the buy-all, sell-all arrangement or *simultaneous purchase and sale contract.*

Electrically speaking, the situations are identical. Electricity flows to the load of least resistance. If a QF sends electricity to the grid and is in need of electricity at that moment, it will be served by its own generator. The only difference is the way the two arrangements are metered. Indeed, one regulatory handbook put out by the New Hampshire Public Service Commission calls the two arrangements different "bookkeeping methods."

However, the two types of contracts can have very different effects. Those QFs that elect net billing may be placed under a tariff distinct from those that purchase all their electricity from the utility. The difference in the revenue that the QF earns is dependent on the utility's charges and PURPA payments. QF owners should analyze utility charges and credits closely before they decide under which contractual arrangement they choose to operate.

To illustrate this, take the example of a typical residential customer who owns a 10-kw wind turbine. The residential consumption is 700 kwh a month. Two capacity factors have been chosen for the wind machine: 15 percent (case A) and 25 percent (case B). Under these capacity factors, the residential wind system generates 1,095 and 2,555 kwh per month, respectively. Seasonal variations have been omitted to simplify the illustration. The final assumption is that the QF could use no more than 50 percent of its own output internally.

Given these assumptions (a real QF owner would make more detailed calculations), the analysis of whether the QF should choose net billing or simultaneous generation and sale is purely an exercise in arithmetic. Those uncomfortable with numbers should work with someone who isn't before making decisions about contracts and approaching the utility.

The first utility used in this example is Green Mountain Power Corporation in Vermont. To evaluate the various cases, first assume the QF sells everything and buys everything. The residence consumes 700 kwh per month. Under Green Mountain's Residential Rate 01 the homeowner pays a customer charge of $5.50 plus a monthly energy charge in peak season (five months of the year) of 300 kwh × 4.07¢ = $12.21 plus 400 kwh × 7.90¢ = $43.81 for a total of $49.31 each month for five months or $246.55 for the peak season.

Once again, to simplify the problem, assume the residence uses the same amount of energy in the off season. It pays the same customer charge of $5.50 plus 300 × 4.07¢ = $12.21 plus 400 × 4.40¢ = $29.81 for a total of $35.31 a month or $247.17 for the off season. The yearly payment is therefore $493.72, and the average monthly energy bill is $41.14.

Under PURPA, Vermont utilities pay QFs 9¢ per kilowatt-hour on-peak and 6.6¢ per kilowatt-hour off-peak. The QF can choose to receive a flat annual rate of 7.8¢ per kilowatt-hour. For this problem, that flat rate is used.

Under the simultaneous purchase and sale contract, where the QF sells everything it generates to the utility under the assumptions of case A, the 10-kw wind turbine will generate income of $85.41 a month or $1,024.92 yearly. This is taxable income. If the taxpayer is in the 30 percent bracket, the after-tax income would be $55.75 a month, or $669.00 a year. Therefore, under a simultaneous purchase and sales contract, the net benefit is $14.61 per month, or $175.32 a year. Under case B, which assumes a higher output from the wind turbine, it earns $199.29 a month or $2,391.48 yearly.

Under a net purchase arrangement the QF has two sources of income: (1) direct sales to the utility and (2) displaced purchases from on-site use.

Since the QF uses 50 percent of the output on-site, it will use 50 percent of 1,095 or 547 kwh per month on-site. The other 547 kwh it generates is sold to the utility. The residence earns 547 × 0.078¢ = $42.67 a month by selling the electricity to the utility, and displaces 400 kwh of peak energy and 147 kwh of off-peak energy a month. In the peak months, the cost of this displaced energy is $37.50 and in the off-peak months it is worth $23.58. The total yearly revenue from the utility is thus equal to 12 × $42.67 = $512.04. This income is taxable. Assuming a 30 percent tax bracket, this is equal to $29.87 a month or $358.44 a year in after-tax income. The QF purchases 547 kwh per month. Based on previous calculations, the QF will pay Green Mountain $34.07 per month or $408.84 per year.

Now look at the various comparisons for case A. Under simultaneous purchase and sale contracts, the QF earns a net benefit of $14.61 per month, or $175.32 per year. In other words, the QF will receive $14.61 more each month from the utility than it will pay to the utility for electricity used on-site. Under a net billing arrangement, the QF will reduce its utility bills, but over the year it will pay out $4.20 more per month or $50.40 per year than it will receive. The difference between the two arrangements to the QF is $18.81 per month. Since simultaneous purchase and sale requires an additional meter, another conclusion of this analysis is that the QF can afford to pay up to $18.81 per month for the meter and still find this type of contract preferable.

Take the same technology and the same residential load but relocate it to the service area of Montana-Dakota Utilities Company. Under its residential electric service Rate 10, the homeowner pays 300 × 5.562¢ = $16.69 + 400 × 6.900¢ = $27.60 or $44.29 per month, or $531.48 a year. There is no customer charge, nor is there seasonal pricing.

If the wind turbine owner signed a PURPA contract for short-term power purchases under their Rate STPP-92, it receives 2.16¢ per kilowatt-hour regardless of the time of year. If it chose a simultaneous purchase and sale contract, it receives 1,025 × 0.0216¢ = $22.14 a month, or $265.68 a year. Given a 30 percent tax bracket, this is equivalent in after-tax income to $15.69 a month or $188.28 a year. The QF pays $28.60 per month ($343.30 a year) more to the utility than it receives. If, on the other hand, it chose a net billing arrangement and used 50 percent of its output to replace its own consumption, it receives 547 kwh × 0.0216 = $11.82 from the utility, equivalent to $8.27 per month after taxes. The QF will purchase $33.90 of electricity each month for internal use ($406.80 per year). Under the net billing arrangement, the QF will pay the utility $25.63 more per month than it receives from the utility.

In this situation a net billing arrangement is only slightly better than the simultaneous purchase and sale contract.

What if this QF signs a long-term contract with Montana-Dakota Utilities Company under Rate LTPP-93? Under this minimum four-year contract, the energy rate more than doubles, and the QF receives a capacity credit based on its capacity factor. In other words, if the wind turbine has a capacity factor of 15 percent, it receives 0.15 ÷ 0.85 ×

$5.33 per kw or 94¢ per kilowatt. For a 10-kw wind turbine, that would equal $9.40 per month. Add to that the 1,025 × 4.38¢ per kilowatt-hour, or $44.90 per month it earns, and it receives a total of $54.30 per month. If the wind turbine capacity factor were to increase to 35 percent, not only would its capacity credit rise to $2.13 × 10 = $21.30 per month, but its energy derived revenue would increase to $117.02, for a total of $138.32 a month.

What emerges from this brief analysis is the following:

1. *If the buyback rate is significantly lower than the purchase price, displaced energy becomes valuable enough to offset the advantages of simultaneous purchases and sales.* This means a net billing arrangement would be preferable. In many states the utility buyback rate is less than half its retail rate. In fact, Ken Hach is developing a switch that automatically diverts excess electricity to a water heater storage tank for those situations where the buyback rate is very low. The electricity brings more value as stored hot water than when it's sold to the utility. Net billing works best for small systems. For systems greater than 10 kw, only a small fraction of the electricity generated could be used internally.

2. *A long-term contract may contain valuable advantages.* It should be noted, however, that the disparity between firm and nonfirm energy in Montana is the greatest in the nation. In most cases, the QF would not receive more than 1¢ to 2¢ more for long-term energy contracts, and if these contracts do not contain escalator clauses for inflation and short-term avoided costs continue to rise, the QF might be better off with an as available energy contract.

3. *The capacity factor of the technology can make a dramatic difference both in the amount of electricity produced and in the ability to qualify for capacity credits.* Once again, Montana's formula for calculating capacity credits is unusual but not unique. That formula allows the QF to receive partial credits based on the ratio of its own capacity factor to that of a baseload coal plant (85 percent). In most states the QF would have to meet a minimum capacity factor (typically 65 to 80 percent). Partial capacity payments are not available.

Remember, the utility may require a customer that selects the net billing contract to come under a different rate schedule than one that buys all its electricity from the utility. This rate schedule may include back-up or standby charges but only if such a rate already exists for that class. The utility cannot impose higher back-up rates for QFs unless it can prove that QFs impose higher system costs. The burden of proof rests with the utility. To prove its claim it must use system cost data. The FERC prohibits the utility from assuming that any outage of the QF facility occurs coincident with the system peak demand. In other words, the utility cannot assume the QF will need electricity at precisely the time the system is at its peak. If it could, it could justify a high back-up charge under the theory that it must have an additional power plant standing by for that time. But if the QF needs electricity when the system has excess capacity, back-up rates should reflect only regular energy charges.

Remember that in bargaining with the utility you are going to have to make trade-offs. In states that establish high buyback rates, you will have more room to maneuver. For example, Dan Darrow, owner of a 10-kw hydro system on a small farm in South Newfane, Vermont, sent a letter in fall of 1981 to the Public Service Board of Vermont and received a copy of the PURPA rates. PURPA rates at the time were 9¢ peak, 6¢ off-peak and 7.8¢ for a 24-hour average. He then wrote to the local co-op, Vermont Electric Co-op, asking them how much they'd give him including in the letter the specs on his machine. Vermont Electric wrote back saying it didn't agree with the PURPA rates, but would offer him 12¢ peak and 3¢ off-peak, with an average of about 5¢ per kilowatt-hour.

Darrow didn't insist on the PURPA rate. Vermont Electric acknowledged that its marginal costs would probably rise sharply, reaching an estimated 18¢ kilowatt-hour by 1986. They were willing to write such projections into a contract. At the time Darrow was interviewed, the final approval on his interconnection equipment and rate was still pending. He knows he could win in a fight over the PURPA rates but would rather wait to see what the utility will offer on interconnection equipment and future rates first.

Force Majeure and Interruption of Power

The QF on a long-term contract wants some guarantee that if its own deliveries are interrupted temporarily because of forces beyond its control it will not be penalized. The QF also wants certainty that the utility will exercise its right to refuse power purchases only at limited times and with sufficient notice.

The first provision is called Force Majeure. The following is a typical paragraph in a contract:

> As used in this agreement, "Force Majeure" means unforeseeable causes beyond the reasonable control of and without the fault or negligence of the party claiming Force Majeure. It shall

include failure or interruption of services due to causes beyond its control, sabotage, strikes, acts of God, drought or accidents not reasonably foreseeable, appropriation or diversion of electricity by rule or order of any governmental authority having jurisdiction thereof, and failure to deliver electricity during such time as it may be obliged to temporarily discontinue delivering the electricity on account of system operating conditions and in the case the service is so interrupted.[7]

Usually utilities require the insertion of the following sentence into a contract: "Force Majeure shall not include the nonavailability of fuel or other motive force to operate Seller's Facility." Utilities argue that without such a provision the wind turbine or hydro operator would be exempted from any capacity penalties if the wind didn't blow or the water flow dropped. The seller, on the other hand, wants some protection against low water flow or low winds and if exempted from Force Majeure status, then it should be specifically inserted into the contract. For example, the seller might not sell electricity during the summer months to a winter-peaking utility.

The QF might also propose a paragraph that allows it to shut down the facility for scheduled maintenance periods. Usually such a paragraph must specify the maximum number of days each calendar period that this can occur (e.g., 30 days). Any payments for capacity shall be conditioned upon scheduling the maintenance of the facility at a time acceptable to the buyer and to the regional power pool.

Although not directly relevant, it might be of interest to note the clause one utility inserted in a contract to a 15-kw wind developer about the utility's responsibility for reliable electric service: New York State Electric and Gas Corporation

will endeavor to provide continuous and reliable service hereunder, but if it is prevented from doing so due to circumstances beyond its reasonable control, or through the ordinary negligence of its representatives, agents, or employees, it shall not be liable therefore.[8]

PURPA allows utilities the right to interrupt power purchases when making such purchases would actually increase their costs. Utilities often try to widen their authority to exercise this right. The legislation as defined by the FERC regulations does not permit the utility to interrupt power just because it would be cheaper for it to buy from someplace else. Most regulatory commissions recognize the right of utilities to interrupt purchases under light load conditions when expensive baseload plants, especially nuclear, are operating at very low capacity. Under such circumstances, purchasing QF power might drop the load carried by the plant to such a low level that the plant has to be shut down. Starting up the plant again would cost more money than running it at this light load. Therefore the QF in this instance imposes negative avoided costs on the system.

The QF should propose a clause for the contract that requires sufficient advance notification of such purchase interruptions and limits these interruptions to a specific maximum time period. PURPA regulations do require sufficient advance notification, so it is the QF's right to have such a clause in the contract. As discussed in an earlier chapter, PG&E limits such periods of nonpurchase to not more than 7 percent of the year.

The QF could try to insert a "take or pay" provision in the contract. This provision might say that "in the event the Seller is prevented from delivering energy because Buyer is unable or unwilling to accept energy or is precluded by the regional power pool or any successor organization from doing so, Buyer shall nevertheless be obligated to make payment." These provisions are common between all-requirements utilities and their bulk suppliers. Utilities that buy a share of a proposed power plant commit themselves to purchase a portion of its electricity even if they find they do not need it at some later time.

Billing and Payment

A minor point, but one that may fulfill some QF owner's fantasy, is to insert a paragraph that requires the utility to pay for electricity within a certain time period. Late payments will be assessed a penalty and/or an interest charge at the current prime rate.

Liability Insurance

The question of how much insurance a QF should have, and what type, is not yet a settled issue. All utilities will insert an indemnity clause in the contract. The following is a typical clause in a contract between a 15-kw wind turbine QF and the New York State Electric and Gas Corporation (NYSE&G):

> *Customer agrees to indemnify and hold harmless NYSE&G, its representatives, agent and employees, from and against any costs, damages, liens, suits, claims, demands and expenses of any kind for any injury or death to any person or any damage to any property caused by or arising from any act of Customer or Customer's representatives, agents of employees in connection with the construction, installation, interconnection, operation, maintenance, disconnec-*

tion or disassembly of any part or the whole of Customer's system, including any change, modification or addition thereto.[9]

Such a clause is acceptable, but only if the seller has the same indemnity protection from the utility. *Indemnity should run from each party to the other for all harm resulting from the establishment, maintenance and operation of the respective equipment of each.*

Most investor-owned utilities do not require specific amounts of insurance coverage. For those that do, the amounts vary, for example, from general liability insurance of $100,000 by Niagara Mohawk Power Corporation in New York to PP&L's requirements in Oregon of $1 million liability (as per their contract) "which limits may be required to be increased by Pacific's giving Seller two years' notice. Such increase shall not exceed fifteen percent (15%) per year."

PP&L also requires property damage insurance, although most utilities do not. Based on the recommendation of the National Rural Electric Cooperative Association (NRECA), most rural electric cooperatives require $1 million in liability insurance. Ken Hach, the Iowa-based supplier of wind turbines who also negotiates PURPA contracts for his clients, believes the QF owner can have this requirement lowered if he or she fights it. Paul Gipe, who advises wind turbine owners in Pennsylvania, points out that NRECA's recommendation of $1 million is three times the normal $300,000 auto liability insurance required by most states. That implies that a small stationary wind generator can do three times more damage than the average driver.

Denver Roseburg, owner of a 10-kw Jacobs interconnected with the McLean Rural Electric Cooperative in Underwood, North Dakota, refused to buy $1 million in blanket liability insurance because it would have cost him $300 per month. He ended up having to swear in an affidavit that he has assets in excess of $1 million.

Co-ops argue that this kind of insurance is actually cheap, only an additional $25 to $50 per year, especially if the homeowner already has business insurance with the same company. Another slant comes from the Gundermeier family, which owns a small, 10-kw wind system in Storm Lake, Iowa, that's interconnected with the Buena Vista Cooperative. Their wind turbine vendor, Hach, told them they could successfully appeal the $1 million liability provision to the Iowa Commerce Commission, but they did not do this. Instead, they paid extra for insurance with the Farm Bureau, their original homeowner policy carrier. Later they switched to State Farm and got a lower overall premium for the entire policy, including the liability insurance. The point is that QF owners should shop around for insurance carriers.

Nonutility Status of QFs

A QF owner may want a paragraph inserted into the contract that reaffirms the nonutility nature of the seller and in most states reinforces the lack of jurisdiction of public utility commissions over the seller's facility. Investors and lenders may insist upon such a provision as protection against governmental intervention. A typical paragraph would read as follows:

> *Nothing in this Agreement shall be construed to create any duty to, any standard of care with reference to, or any liability to any person not a party to this Agreement. No undertaking by one party to the other under any provision of this Agreement shall constitute the dedication of that party's system or any portion thereof to the other party or to the public, nor affect the status of Buyer as an independent public utility corporation, or Seller as an independent individual or entity.* [10]

Resolution of Disputes

The QF can appeal to the regulatory commission before a contract is signed. Disputes that occur after the contract is in effect might best, from the QF's perspective, be submitted to binding arbitration by an independent arbitrator mutually agreed upon by the parties. The arbitrator should be empowered to make binding final awards including monetary damages and specific performance.

Binding arbitration is a relatively quick way to settle disputes. This can be important if during the dispute the utility is purchasing none of the QF's power. Such an interruption could easily bankrupt the QF if allowed to continue too long.

Contract Modification and Termination

A short-term contract usually runs for a year. Either party can give the other relatively brief notice for termination or contract modification. At least one New York State utility has offered a one-year contract to a QF that could be terminated "by either party upon 48 hours prior written notice to the other." On the other hand, long-term contracts longer than one year often contain termination or modification provisions that are very long. In some cases the QF could not modify or terminate the contract in less than two years. Thus QF owners who found that their generators operated at a higher than predicted capacity factor, or who wanted to switch to a time-of-day sales arrangement or from net billing to simultaneous purchase and sale, would find themselves stopped from doing so for a significant time period. QF owners must make the decision as to whether they want to give themselves sufficient leeway to modify the contract, knowing that the utility can do the same, or whether they feel the contract is sufficiently favorable to tie the provisions to a long advance notice provision for modification or termination.

Interconnection Requirements

The area of interconnection requirements is the thorniest of all that the QF owner will encounter. (See chapter 4 for a discussion of the technical aspects of interconnection.) The FERC gives no guidance on interconnection standards. However, it did say that "interconnection costs of a facility which is already interconnected with the utility for purposes of sales are limited to any additional expenses incurred by the utility to permit purchases." [11] The vast majority of regulatory commissions have been reluctant to become involved in this area, leaving interconnection standards to the individual utilities.

This is one question that is still very much in a legal limbo. The Federal Power Act specifically states that utilities do not have to interconnect if doing so would undermine the reliability or safety of the grid. Until PURPA was enacted anyone wanting to interconnect had to apply to the FERC. The utility could deny the application, and the decision by the FERC would be based on the individual merits of that case. In attempting to follow the spirit of PURPA, the FERC cut through the red tape process. It basically decided that QFs would not undermine the reliability or safety of the grid system and required utilities to interconnect with them. Utilities would prefer to have the option of requiring a case-by-case evaluation before the FERC. That type of evaluation would add significantly to the time and expense of interconnection and, if required of very small QFs, would simply negate the intent of PURPA.

In any event, most utilities are treating each QF on a case-by-case basis. All states require the utility actually to do the installation or to oversee it. All interconnection arrangements must meet with the approval of utilities. No QF is permitted to come on-line without the utility's inspection and approval of the interconnection equipment.

Those utilities that have developed standards often differentiate among sizes of QFs. This is as true within one state as it is between states. PG&E has interconnection requirements for QFs of up to 20 kw, for those 20 kw to 100 kw, and for those over 100 kw. Southern California Edison has interconnection requirements for those under and over 200 kw. In New York State, Central Hudson Gas and Electric

Corporation has a general interconnection standard only for wind turbines of under 15 kw. All the rest must be treated on a case-by-case basis. Niagara Mohawk Power Corporation limits its general standards to QFs of under 50 kw. NYSE&G has separate requirements for systems greater than 15 kw and less than 15 kw. Rochester Gas and Electric Corporation (RG&E) requirements apply to all systems of 80 Mw or less. Orange and Rockland Utilities, Inc., has general standards only for those under 10 kw. RG&E doesn't allow QFs with greater than 45 kilovolt-amperes (kva) to connect to its low voltage lines. Consolidated Edison (Con Ed) doesn't even allow QFs with greater than 10-kw capacity to connect to its low voltage lines.

Sometimes utilities will use the interconnection standard to discourage QFs. That may be difficult to prove, but most utility engineers admit they are extremely conservative about interconnection standards. They dream up any combination of circumstances and faults and then require redundant systems to protect against such possibilities. Pennsylvania Power and Light interconnection standards are of this type.

Ed Zimmerman, president of E-Z Manufacturing Company, installed a 125-kw gasifier induction generation system in a factory that produces charcoal as well as sawmill equipment. Pennsylvania Power and Light pays him an attractive 6¢ a kilowatt-hour buyback rate. According to Zimmerman, it took a full year after he first tried to go on-line to complete negotiations about interconnection hardware requirements. Pennsylvania Power and Light requires over/under voltage/frequency/current, plus phase unbalance and individual phase monitoring. Zimmerman had to include a mechanical relay on top of an electronic relay for each function. An electronic voltage relay costs $150, and a mechanical relay costs from $500 to $600. Plus Pennsylvania Power and Light forced him to pay $30,000 to replace the hydraulic overcurrent relays (two) then serving his step-down transformer with electronics in order to guarantee that the company's circuit breaker would not reclose in the event of fault conditions on that line that continued over a longer-than-normal time frame. This prevents the theoretical possibility that Pennsylvania Power and Light capacitors, installed for voltage compensation "at the end of the line," would allow the induction generator to continue operating, a circumstance which Zimmerman protests is "impossible, not only because line voltage would collapse but even if it happened to end up in that range of my over/under voltage relays, my frequency relays would trip the breaker and shut down." He fought bitterly against this additional cost but finally acquiesced because he was losing money by not being able to interconnect. In his case the total interconnection cost was $60,000, more than a third of the total cost of the installation.

Ted Keck, owner of a 70-kw hydro site in Pillow, Pennsylvania, says that Pennsylvania Power and Light required him to install over/under voltage relays on each of his three phases, plus over/under frequency on one phase for a total of 8 relays costing from $300 to $600 each. Pennsylvania Power and Light requires utility-grade relays, which are more expensive than industrial-grade. Keck at first thought this was excessive but now allows it is "probably for my own good" and notes that utility-grade relays last for 40 years, almost ten times longer than industrial-grade. Pennsylvania Power and Light insisted that he buy two transformers from them for $500 each, which he probably would not have used. He expects that the total, installed cost of interconnection equipment will be $4,000 to $5,000.

The easiest interconnection we've come across took place in Connecticut. Connecticut Light and Power (CL&P) connected to a 12-kw hydro site owned by a firm of solar architects and a 60-kw site owned by a factory making hi-tech plastic laminates. Bill Johnson, president of Newfound Power Company in Rhode Island (a firm that publishes *Currents,* a hydro quarterly, and develops sites) did the negotiations. At the 12-kw site he's using only a "magnetic starter," made by Alan Bradley Company. He describes the device as a "heavy-duty electro-magnetic industrial relay." It appears to be a circuit breaker with a built-in overcurrent relay. No other relaying or dedicated transformer was necessary. This case also illustrates the need to discuss the electrical equipment intelligently.

CL&P originally wanted three phase with over/under voltage relays simultaneous with phase balance monitoring. However, they were persuaded to copy the requirements of Narragansett Electric, another utility with which Johnson was negotiating. It took Johnson one month to work out the interconnection arrangements with Narragansett and two months with CL&P.

One would expect that the rate of innovation in the microprocessor industry would soon bring into the marketplace "black box" devices that perform the protection functions more cheaply. That is already happening. The Beckwith Electric Company of Largo, Florida, has developed a solid-state device that handles all eight relay functions for a cost under $2,000. Pennsylvania Power and Light is testing the box and hopes to designate it utility grade for use in its program.

The advances in interconnection equipment may be coming along fast but some utilities impose standards that make it difficult to integrate these advances rapidly. For example, Long Island Lighting Company (LILCO) and Orange and Rockland Utilities require Underwriters'

Laboratories (UL) approval for all equipment. The problem with that requirement is that UL approval is based on a certain amount of operating experience. Thus equipment like Beckwith's would not gain approval for some time. UL approval as of mid-1982 restricts one's search to General Electric and Westinghouse equipment, built for utility companies or very large industrial cogenerators.

Zimmerman raises an oft-repeated complaint by QFs. Utility engineers are not used to induction generators. So they design standards for synchronous generators and refuse to change them. Induction generators, as the previous chapter discussed, can operate only when the outside line is energized, because they need to be excited by grid electricity (unless a bank of capacitors is located nearby). Moreover, they are in phase with the grid for this same reason. Synchronous generators can operate in isolation with the grid and, therefore, need a great deal of additional synchronizing equipment and relay protection.

One sticky problem is that the utility and the QF owner might agree on specific wiring diagrams but the local building inspectors might prohibit them. Some utilities require that electrical wiring adhere to national codes, others to state codes and still others leave that open. Thus the QF must work closely with the local electrical code official as well as the utility engineer and the QF's engineers.

Many utilities require QFs to supply either single or three-phase current, while others allow both. In New York State, for example, NYSE&G allows single or three-phase. RG&E allows single or three-phase if the generator capacity is less than 48 kilowatt-amperes (kva). If larger than that, it must be three-phase. Central Hudson Gas and Electric requires single-phase current. Niagara Mohawk has no specifications. LILCO requires three-phase circuits.

The vast majority of utilities provide general requirements about the quality of electricity supplied by the QF. Some are quite specific. Central Hudson Gas and Electric requires an oscillographic print showing the wave shape of the voltage or current supplied to the network systems at the interconnection output terminals. RG&E not only requires a similar print but wants two, under "light load and full load conditions." The print must include current magnitude, harmonic content, power output, and estimated or measured power factor of generator load.

Some utilities have no harmonic standards, although most do. Niagara Mohawk requires total harmonic distortion (THD) of not more than 5 percent, but adds that in certain cases a more stringent limitation may apply. Central Hudson Gas and Electric requires a THD of less than 10 percent of the fundamental voltage measured at the point of delivery. NYSE&G allows a maximum distortion of 3 percent for any single frequency or 5 percent THD.

It is unclear what recourse QFs have to onerous interconnection standards. The FERC regulations make clear that these standards should not be used to discourage QFs. Also, QFs can appeal to the regulatory commission if they feel that a double standard is being used. Thus, for example, the Potomac Electric Power Commission (PEPCO) argued before the Washington, D.C., Public Service Commission that it should not be required to interconnect QFs of greater than 300 kw to its low voltage lines because of the burden it would place on the transformer and distribution lines. The electrical engineer testifying for QFs countered that there were several customers with loads of more than 300 kw already on the line and that the existing transformer had several megawatts of capacity. The customer load forces that amount of electricity to be supplied through the line and transformer. Therefore PEPCO should not discriminate against a QF that imposed similar demands on the existing distribution system.

Similarly, utilities that have no standard for harmonic distortion for their nonproducing customers should arguably not have one for QFs. Steve Strong, of Solar Design Associates in Lincoln, Massachusetts, has designed two residential photovoltaic installations in Massachusetts and New Mexico. He comments, "If you measured the waveform on any utility distribution feeder, you'd find lots of 'hash' — bad power factor, overvoltage and harmonic distortion caused by interference by neighboring loads on the line" especially during full-load periods.

Angelo Skalafuris, vice-president of Re-Energy Systems, which sells small cogeneration systems, goes a step further. The waveform coming out of a central utility plant may be almost perfect. But even an infinitesimal imperfection is magnified when the signal passes through a transformer. The current will pass through two and possibly three transformers on the way to its final destination. Each time, the slightest out-of-phase tendency will be vastly exaggerated. Skalafuris has monitored a 100-kw gas turbine cogeneration system at a hotel in Philadelphia and found that the meter was actually running faster when using the utility's electricity. A significant reason in Skalafuris' estimation was the "dirtier" electricity coming in from outside.

If utilities already have standards for harmonics or for independent power production, it is more difficult for QFs to argue for exemption even though the existing standards were designed for the utility itself or for large industrial cogenerators. Numerous studies have been done to indicate that small QFs, especially those of under 100 kw, put no burden on overall system reliability even at high penetration levels and that small numbers of QFs with capacities of less than 20 kw need not have high-quality protective devices.

Utilities often require a dedicated or isolation transformer. Yet this is not always the case. Steve Strong notes that neither Boston Edison nor Public Service Company of New Mexico required a dedicated transformer for the 7.3-kw and 3-kw respective photovoltaic arrays.

Paying Interconnection Costs

The cost of interconnecting with the grid system can be high. Several QFs interviewed indicated payments that were up to one-third their total cost of installation. Most regulatory commissions allow utilities to finance such costs. Utilities, at their discretion, can allow the QF to repay the costs over several months or years. Several states require utilities to provide for monthly payments to amortize the interconnection costs. In such states the repayment period is often based on the term of the contract. It might be a percentage of the term of the contract, for example. Even in these states, however, the utility need not amortize the interconnection costs if the QF is not credit worthy. In cases where the utility extends the QF credit, it charges interest, usually at the prime rate.

Most utilities require full payment of interconnection costs up front, prior to the commencement of service. In some cases, the interconnection equipment is owned by the customer if paid for by the customer. In most instances, it is owned by the company even if paid for by the customer.

Some utilities, usually those that try to impose a very high meter charge (e.g., $50 per month) try to impose additional interconnection expenses. Con Ed, for example, proposed to the New York State Public Service Commission that, over and above its full cost reimbursement for interconnection costs, "the Customer shall pay an annual charge of 9 percent of the capital costs of interconnection, including the costs of distribution system reinforcements, to cover property taxes and operation and maintenance expenses." The QF can and should require the utility to justify this additional expense. Such justification would include submission of its past property tax payments as well as its *additional* operation and maintenance expenses for the QF interconnection equipment.

The field of interconnections is probably among the most rapidly evolving of all those the QF will encounter. New technologies that perform protective functions more cheaply and reliably than existing technologies will be entering the marketplace. As greater numbers of

QFs come on-line, their impact on the secondary distribution system will change. One might expect that interconnection standards would evolve for different-size technologies and for different types of grid systems (e.g., radial or network) and for different types of generators (e.g., induction or synchronous). To repeat the basic advice of this chapter: *Those who are operating or want to operate under PURPA must establish city-wide, statewide and national associations to deal with these issues.*

So far the actual types of equipment and systems available to the homeowner or small business person desiring to enter the field of small power production have yet to discussed. The following chapter covers four basic systems: wind turbines, hydroelectric plants, photovoltaic systems and cogeneration plants.

CHAPTER 6
Electric Generation Technologies

The Public Utility Regulatory Policies Act (PURPA) has created a new market for electricity produced by independent, small-scale generating systems. Homeowners, landowners and businesses wanting to generate power on-site no longer have to store the surplus in expensive and inefficient battery systems. These small-scale systems can be connected to the grid to sell electricity profitably and buy back-up power economically. Engineers, inventors and others have designed a variety of small power plants to meet the newly created demand.

The first generation of modern, dispersed power plants made its debut in the latter part of the seventies. By 1982 a second generation appeared, with significant refinements. New designs and materials have been coupled with sophisticated electric technology to increase the efficiency and reliability of the newer systems.

Production facilities have outgrown backyard garages and moved into suburban and rural factories. Higher production runs have lowered equipment costs. Spin-off companies have begun to provide related services, such as site evaluation, servicing and financing. The industry has definitely grown out of its infancy, a little wobbly at first, but then with increasing confidence. Now it walks firmly on two legs.

This chapter surveys the four electric generation technologies that are available for the home and for small commercial enterprises: wind power, hydropower, photovoltaics and fossil-fueled cogeneration plants. The purpose here is not to provide a how-to lesson for building and installing on-site power plants but to discuss the ways small power-producing technologies work and how to estimate the power output one will get from various systems. Detailed books on each of these technologies are available to provide how-to information. Here the objective is more modest: to inform the prospective buyer on how these systems work, how to assess their appropriateness for a particular site and how to choose the proper equipment.

Each of these four technologies has its own unique characteristics and its own jargon. Wind developers speak of power density and

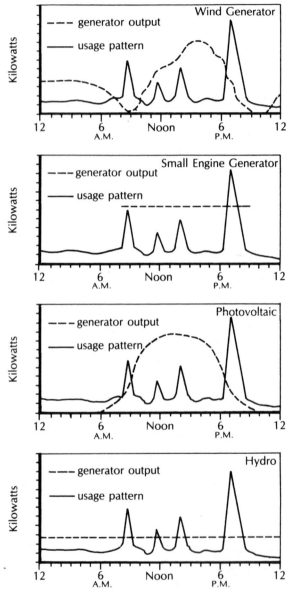

Figure 6–1: Notice the sharp spikes for individual residential loads, compared to the diversified load curves seen in previous chapters (see figure 2–2). The loads are the same, but the timing of the electrical output by different systems varies dramatically. Hydro and small engine generation provide the best match. *Redrawn from Terrance D. Paul,* How to Design an Independent Power System, *copyright © 1981, with permission of Best Energy Systems for Tomorrow, Necedah, Wis.*

Rayleigh distributions. Hydro developers talk about feet of head and flow duration curves. Photovoltaics suppliers discuss peak watts and junction barriers. Cogeneration manufacturers discuss heat exchangers, Carnot cycles and Otto cycles.

Each operates on different principles. Wind and hydro machines convert the force of moving air and water molecules into mechanical energy in a turbine and then use the turbine to generate electricity in a generator. Cogenerators use the internal combustion engine to rotate a turbine that turns a generator to produce electricity. Photovoltaic devices have no turbines at all. They capture the energy of tiny bits of sunlight, called *photons*, to create a flow of electrons, an electrical current.

Cogeneration systems generate useful heat as well as electricity (or mechanical power), using the same primary fuel. The concept of cogeneration can be applied to photovoltaics. Silicon-based photovoltaics convert at most about 20 percent of the sunlight striking the cell into electricity. The other 80 percent becomes waste heat, which could be captured if the means to do so become less expensive than the value of the thermal energy captured. In fact, silicon-based photovoltaics are more efficient if they are cooler. Thus, by removing heat by moving air or water over the cells, the electrical output increases as the thermal energy is harnessed.

All four technologies strive to achieve the same objective: to convert the largest portion of potential energy into useful work for the least cost, or to have the highest efficiency and the lowest price per unit of energy produced.

All of the technologies except photovoltaics exhibit economies of scale, meaning the larger the power plant, the cheaper the cost per kilowatt of power produced. Economies of scale are not unlimited. Beyond a certain scale, the increased structural stress on the materials in the wind turbine may outweigh the advantages of large blade diameters. The reliability of several smaller cogeneration systems is greater than that of one large system and can spell the difference between a cost-effective and an unprofitable arrangement.

The most cost-effective system is one sized to fit the available fuel supply and demand. A wind turbine built to function effectively in high-velocity winds will often generate electricity inefficiently in low-velocity winds. A hydroelectric plant sized to the maximum seasonal flow rate will operate very inefficiently during the majority of the year when flow rates are lower. A cogeneration facility sized to satisfy a winter heating load will generate too much heat in the summer months.

Many designers of cogeneration systems avoid the problem of excess waste heat by sizing the facility to meet the annual base load for thermal energy, often the domestic hot water load. If the PURPA buy-back rates are high, the cogeneration system will be sized much larger, in order to maximize its electrical output. A great deal of work is now going on in the industry to allow the facility to use the excess heat resulting from this larger-size facility. Some companies are linking absorption air conditioners to their cogeneration systems. The high-temperature waste heat drives the air conditioner in the summer months.

The best advice to anyone entering the field of small power pro-duction is also the most obvious: piggyback on the experience of others. Learn from the first- and second-generation entrepreneurs. By late 1982 more than 500 PURPA contracts had been signed with small-scale cogenerators and small power producers. Each contractor often owned more than one power plant, and more than a thousand turbines and generators of all sizes were in operation. Chances are that within 100 miles of where you are there are at least two pioneers selling electricity to the utility. Some are wary of visitors. The mass media publicized their ventures prematurely, and swarms of curious on-lookers trespassed on their property or otherwise invaded their privacy. Most will respond warmly and openly to someone who demonstrates a working knowledge of small power plants. Don't expect a small power producer to give you a history of PURPA. Do your basic home-work before you visit. Familiarize yourself as fully as you can with the technology you plan to employ.

Your time with the owner will be best spent finding out about actual experiences with the technology and with the utility. The lessons he's learned can save you time, trouble and money.

Wind Power

The Equipment

A hundred years ago millions of windmills dotted the landscapes of many countries, providing mechanical power for threshing grain or pumping water. These mechanical functions are still economically performed by windmills. This book, however, focuses on twentieth-century wind systems that generate electricity. Wind-electric machines are built somewhat differently from their water-pumping ancestors.

The blades or *rotor* of a wind machine convert the wind's energy into mechanical or rotary power. Modern wind machines are designed to gain an aerodynamic lift from the wind in much the same way that airplanes do. The individual rotor blades of most modern wind genera-

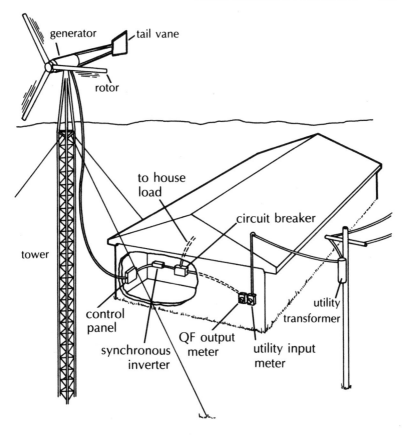

generator
tail vane
rotor
tower
to house load
circuit breaker
control panel
synchronous inverter
QF output meter
utility input meter
utility transformer

Utility-Connected System

Figure 6-2: Notice the two meters in this grid-connected wind electric system. One monitors consumption from the utility transformer and the other output to the utility grid.

tors have an airfoil cross section. When the wind hits the blades, the resulting aerodynamic lift causes the blades to move faster than the wind itself. Advances in wind machine blade design have helped to optimize the efficiency with which these systems extract energy from the wind. The relationship of the rotor's speed (measured at the blade tips) to the wind speed is the *tip-speed ratio*. If the blades are moving five times faster than the wind, the tip-speed ratio is 5:1. Windmills with low ratios of about 1:1 are mainly suited for slow-speed, mechanical purposes such as water pumping. High-tip-speed propellers in the

Inside a Windcharger

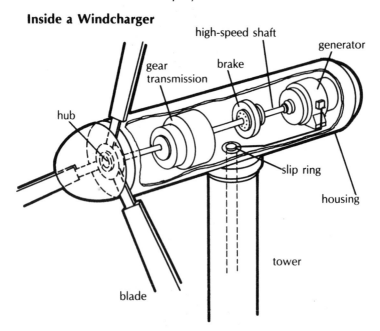

Figure 6–3: The wind drives the blades that rotate the shaft inside a windcharger.

range of four to eight times faster than the wind are suitable for generating electricity.

Wind machines are classified according to the orientation of their axis of rotation. They are either *horizontal-axis* or *vertical-axis* machines. At present horizontal-axis machines greatly outnumber vertical-axis designs. The chief advantage of the vertical-axis wind machine is that it is easily able to accept wind from any direction.

Horizontal-axis machines can be further sub-divided into upwind or downwind machines. *Upwind rotors* are steered into the wind by a tail vane. Advocates of the upwind design point out that it avoids the "wind shadow" caused by the tower and therefore operates in a more even wind flow.

Downwind machines don't need a tail vane. Some designers, such as Hans Meyer, founder of Windworks in Mukwonago, Wisconsin, believe this is a more economical design for larger machines. He points out that, as the rotor diameter increases, "a tail structure that puts the tail sufficiently clear of the slipstream (the flow of wind) becomes a big and expensive structure."[1]

Besides the rotor or blades, the typical wind machine has a *governor* or controller. Different manufacturers use different methods of

Photo 6-1: The Darrieus rotor pictured here is a vertical-axis wind machine that can be driven by wind from any direction. The transmission and generator are located at the base of the vertical mast. *Photograph courtesy of Joe Carter.*

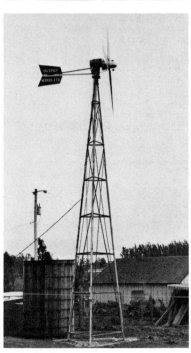

Photo 6-2: An upwind horizontal-axis machine uses a tail vane to orient the rotor to face into the wind. *Photograph courtesy of Joe Carter.*

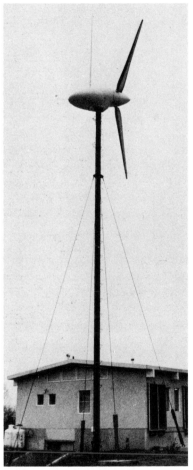

Photo 6-3: A downwind horizontal-axis machine operates without a tail vane. The rotor itself is oriented downwind of the tower mast just by the force of the wind. In this photograph the wind is coming from the left. *Photograph courtesy of Joe Carter.*

blade-pitch control and overspeed protection. Turbines with small rotors often use no pitch control or rely on flexing blades to conform to variations in wind speeds. Larger machines, however, require some form of governing to adjust the blade pitch for both efficiency and speed control. Often there is a trade-off between simplicity and increased efficiency in larger machines. As Karl Bergey of Bergey Windpower points out, "High aerodynamic efficiencies are obtained with... variable-pitch rotor systems at the cost of mechanical complexity and decreased reliability."[2]

Field data from wind machines supports the view that the most economical machine is the one that can survive infrequent but very high wind gusts. A 100-miles-per-hour (MPH) gust can accelerate the rotor tip from a moderate speed of 175 MPH in a 25-MPH wind to 300 MPH to 500 MPH. When this happens, the machine is almost certain to be damaged unless it has some form of overspeed control. The simplest method is to put a brake on the rotor and run a cable to the ground. In high winds the owner can use the cable to brake the rotor. However, since wind speeds can increase very rapidly, the operator needs to be nearby for this manual system to work well.

Most of the upwind generators have some form of pivoting tail to swing the rotor out of the direct wind at about 30 MPH. This is called yaw orientation. One method allows the rotor and generator assembly to tilt up, out of the wind. Another method positions the rotor so it keeps turning, but at a reduced power output. Some maintain a reduced power output by using a blade pitch control that turns the blade to a pitch angle of 60 degrees. This feature is commonly used in horizontal-axis designs, in upwind and downwind configurations.

Gears, like the transmission in a car, are used to increase the turbine's speed of revolution where it connects with the generator. (Gears also reduce the overall efficiency of the system because of energy losses due to friction.) Most systems have a transmission to get the shaft speed up to the 1,800 revolutions per minute (RPM) of a typical generator. Some wind machines have no gears. These direct-drive designs use special, low-speed alternators that are coupled directly to the rotor shaft.

About half the wind machines sold today can be had with induction generators, the advantages of which were discussed in chapter 4. The primary advantage is that since the magnetizing (or excitation) power for an induction generator comes from the utility line, the generator output automatically matches the frequency and waveform of the utility. This obviates the need for expensive synchronizing equipment. If the line power fails, the generator goes dead, thereby reducing the need for protective relays.

Another advantage of induction generators is their price. A 25-kilowatt (kw) induction generator costs around $450, whereas a 25-kw synchronous generator, sometimes called an alternator, costs $2,000.

Paul Gipe, a technical adviser to wind developers in Pennsylvania and an author of several how-to manuals, offers another advantage of induction generators. Because the generator is tied to the utility power, it maintains a constant rotor rotation at all wind speeds up to the capacity of the generator, at which point the generator fails and the rotor is unloaded. Thus blade-pitch controls aren't needed to obtain a constant rotor RPM, as they are for synchronous generators. Pitch control is needed to prevent overspeed but Gipe says induction control "can be more easily incorporated into a rotor than can an infinitely variable pitch-changing mechanism."[3]

Not all wind machine manufacturers agree, however, that constant rotation is an advantage. Since induction generators are locked into a fixed rotational speed by the power line, they "are usually running too fast or too slowly and do not fully harness the wind energy available," says Marcellus Jacobs, president of Jacobs Wind Electric Company. Jacobs' own alternating current (AC) machine delivers 240 volts (v) through an inverter system, which allows the blade speed to vary with wind speed.[4]

Another disadvantage of the induction generator is that it can act like a motor if the wind speed is too low to spin the turbine fast enough. If this occurs, it will consume rather than produce electricity. The induction generator is essentially identical to an induction motor. The rotational speed of the shaft defines whether it converts electrical energy into mechanical energy or mechanical energy into electrical energy. If rotation of the shaft increases from 1,740 RPM to 1,800 RPM, the motor generates 60 hertz (Hz) electricity. To prevent the generator from acting like a motor, some manufacturers have installed a control box relay that directly couples the generator to utility power only when the wind speed is 10 MPH or greater. This system guarantees that there will be enough wind to drive the induction machine fast enough to produce, not consume, power.

A final disadvantage of the induction generator is its relatively low power factor. Remember, the power factor is an indication of the amount of reactive power consumed by the generator. Reactive power is parasitic. It does no useful work as it circulates through utility lines, but it costs the utility money to supply it. Reactive power exists when the voltage and current waveforms are out of phase, when the peak voltage occurs ahead of the peak current (lagging power factor) or behind it (leading power factor). This out-of-phase condition, or more specifically, the lagging power factor, is an inevitable aspect of induction generations that adds to the utilities' transmission costs.

Some utilities impose penalty charges for poor power factors. Southern California Edison, for example, imposes a 20¢ per kilovar penalty charge when the power factor is below 80 percent. Properly sized capacitor banks can be used to shift the current and voltage into phase with each other to ensure that the minimum acceptable power factor is maintained. However, installing such capacitor banks raises another problem. When the circuit is de-energized, the electricity in the capacitors could be transmitted to the induction generator, thus allowing it to continue to function, eliminating the inherent safety features of a generator that depends on utility electricity for its operation.

It was mentioned before that to prevent damage to the rotor and tower, every wind machine has some form of overspeed protection. This is also used to protect the generators which cannot produce more power than their rated capacity without being damaged. The speed at which the generator produces its rated power is called the *rated wind*

Simple AC Generator

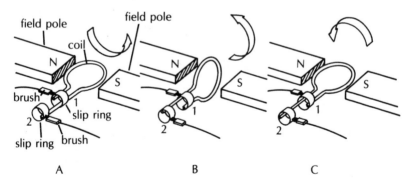

A   B   C

Figure 6–4: In an AC generator, as in a DC generator, the current is always flowing in the same direction in the coil when the coil is perpendicular to the field pole. In an AC generator, unlike the DC generator, copper rings called slip rings are used. Carbon brushes make continuous contact with the same respective slip ring, but as the rings rotate, the current reverses. In this illustration current first flows from slip ring 2 to slip ring 1 and then flows from slip ring 1 to slip ring 2. The result is that the current changes direction with each cycle, or in this example, a 180-degree rotation of the coil. Hence we have the term *alternating current. Adapted from* Wind Power for the Homeowner © *1981 by Donald Marier. Permission granted by Rodale Press, Inc., Emmaus, PA 18049.*

Questions to Ask a Manufacturer

1. How long have you been in business?
2. How well have your products performed?
3. Does your system have a reliable safety system?
4. Is data available from machines at sites similar to mine?
5. Are there owners of this specific machine to whom a prospective customer can talk?
6. What are the skills needed to maintain a particular machine?
7. What type of warranty does the manufacturer provide? What are the warranty limits? Can the owner fix any part of the machine without affecting the warranty?

Questions to Ask a Dealer

1. How many machines have you installed?
2. Can prospective buyers talk to previous customers?
3. How much experience and training in servicing the machines do you have?
4. What constitutes an adequate site analysis and how long will it take (e.g., hours, days, weeks or months) to have a wind system installed, serviced or repaired if new parts are needed?
5. How long is the test period before the owner takes title?
6. What is the role of the dealer in honoring warranties and guarantees?

SOURCE: Reprinted and adapted from *The Kansas Wind Energy Handbook,* Kansas Energy Office, 1981.

speed. For winds that exceed the rated speed, commercial systems are usually designed so the rotor spills some of the wind (overspeed protection: brakes, blade-pitch control, tail vane control); output power thus remains at roughly the rated capacity, preventing damage to the generator. For winds above the cut-out (shut down) speed, the machine must be completely shut down to prevent mechanical or electrical damage.

Finding the Wind Power at Your Site

The availability of wind power, like any of the solar-renewable energies, is highly site specific. Solar energy can be measured and averaged for a relatively large region. The power available in falling water can be pretty well pinpointed with stable stream conditions (the main variable being the amount of snow and rainfall upstream). Wind power can also be measured and averaged, but it is much more capricious. A 10-MPH average for a given region doesn't necessarily mean a 10-MPH average for a site in that region. It could be more or less, depending on such local factors as the shape of the land in and around the site and larger "macro" factors such as weather patterns for a given season or year. You can't simply assume that a site has adequate wind power just because it "seems" that way. You have to study and measure to find just how much wind power is available.

In this regard it's important to know how power varies with wind speed and what kind of efficiencies are possible in converting available power to electric power. The primary rule of thumb is that the power available in the wind varies with the cube of the wind speed. This is a rather dynamic relationship because, for example, with just a 10 percent increase in average wind speed (say from 10 MPH to 11 MPH) there is a 33 percent increase in available wind power. "Available" means that a given wind power system has that much more "fuel" to convert to electricity. On average, wind systems convert between 16 and 30 percent of the available wind power into usable power. All these factors point to the critical importance of making a *site analysis*.

You can start by collecting existing wind speed data from local weather stations and airports. The U.S. Department of Energy has commissioned detailed studies of wind "regimes" throughout the country. This kind of information is only a beginning that can tell you if your site is in the ballpark of wind power feasibility. The next step is to use anemometers to measure, record and average the wind speed at your site. There are a number of types available for rental or purchase, each with varying degrees of sophistication in what they do with raw wind speed data. The most sophisticated can tell you all you need to know, while others will require you to "boil down" a lot of data manually to get meaningful results.

Average wind speed is but one of several important factors in making assessment of the abundance of your wind resource. You have to get "inside" that average to know how often winds are faster than the average. For example, because of the cube law, with 5 hours of 20-MPH wind and 5 hours of 10-MPH wind (averaging up to 15 MPH),

Photo 6 – 4: A simple but important wind-prospecting tool is the anemometer. Connected with a recording device, it can help you collect data about wind speed and direction when making a wind site assessment. *Photograph courtesy of John Hamel.*

there is almost three times more available wind power than there is with 10 hours of 15-MPH wind. Along with daily variations, there are also seasonal variations in average wind speed. There are variations due to the terrain of the site and the height of the anemometer relative to the planned height of the wind machine. If you're looking hard at wind power as a PURPA-type investment, you'll ultimately want your data to tell you just how many kilowatt-hours of electric power different wind machines will produce at your site. Fortunately there is data collection technology and published methodologies, and there are professional concerns that can assist you. The list of publications in Appendix 6 is a good place to start. If you have an idea of the average speed at your site, refer to figure 6–5 for an idea of just what wind machines with different rated outputs can deliver.

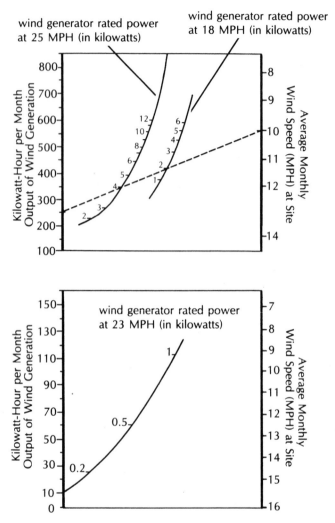

Figure 6 – 5: These nomographs provide estimates of the monthly kilowatt-hour production of wind machines with different power and wind speed ratings operating in different wind regimes. There are three rated wind speeds: 25 MPH, 18 MPH and 23 MPH. Power ratings are 2 kw to 12 kw at 25 MPH, 1 kw to 6 kw at 18 MPH and 0.2 kw to 1 kw at 23 MPH. The dotted line in the upper nomograph shows how to derive an estimate. The machine in question is rated to produce 4 kw in a 25-MPH wind, or 2 kw in an 18-MPH wind. When operated at a site with a 10-MPH average wind speed, the monthly power production is about 250 kwh. Average wind speeds change from season to season, so if you happen to know your site's wind speed average on a month-by-month basis, you can make a more precise estimate of annual power production.

Keeping the Wind Turbine High

Tower height is a very important factor in the ultimate output of a system. Wind speed increases at higher altitudes, in part because the air moves faster and in part because the wind machine is removed from turbulent air near the ground (where wind is slowed by surface friction). The rough rule of thumb is that wind speed varies by the one-seventh power of the height above the earth's surface. Double the height, and the wind speed increases by the one-seventh power. The cube rule translates this into a profound difference in output. For example, at 10 MPH, a Dunlite 2,000-watt (w) wind system will produce only 280 w; at 15 MPH the same wind system will produce almost 1,000 w, almost four times as much. This leads us to another basic law of wind machines: *The least expensive way to get more power from the wind is to build a high tower*. You can choose the most economical tower height by comparing the cost of the last 10-foot increment of tower with the percentage increase of power you might get from the wind machine. Since the tower is often the least expensive component of a wind system, the height might not be restricted by economic considerations but by zoning provisions or structural requirements.

Commercial towers are built of welded triangular design of high-strength steel tubing and are hot-dipped galvanized after fabrication. In some designs each section is 10 feet long, and all sections simply plug into each another. Two types of towers are available: the self-supporting tower and the guyed tower. Self-supporting towers are about 80 percent more expensive (for example, costing around $2,700 for a 60-foot self-supporting tower, compared to $1,600 for a 60-foot guyed tower). They do, however, take up less ground space. Guyed towers have cables that extend outward, requiring a horizontal distance from ground anchor to base of almost 80 percent of the tower's height. Hence the space required for a guyed tower is much greater.

Another reason for putting a wind plant high is to keep it away from disturbed air. The machine should be a minimum of 50 to 60 feet aboveground and in all cases, the propeller tips must be 25 to 30 feet above all obstacles within a 300-foot radius.

The relationship of the height of the tower to the wind speed will also vary by the nature of the terrain surrounding it. Wind moving over the surface of the earth encounters friction caused by the turbulent flow over and around mountains, hills, trees, buildings and other obstructions in its path. These frictional effects decrease with increasing altitude above the surface. Frictional effects differ from one surface to another depending upon the roughness of the surface. Likewise, the

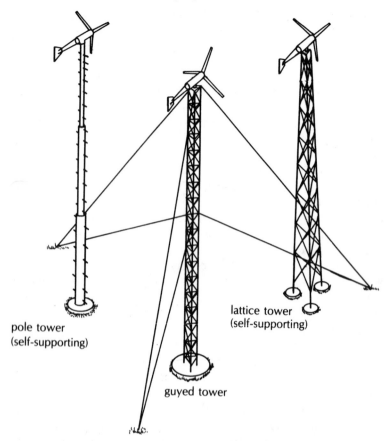

pole tower
(self-supporting)

lattice tower
(self-supporting)

guyed tower

Figure 6 – 6:   Towers for wind systems are either self-supporting or secured by guy wires. Notice the land area needed for the wires holding the guyed tower.

rate at which wind speed increases with height varies with the degree of surface roughness. Wind speeds on higher towers increase at the greatest rate over rough terrain, such as that in the Appalachian Mountains and at the least rate over smooth terrain like that of the Great Plains. Because of this, it is more important to use a tall tower when siting in hilly terrain than, say, in the flat Texas Panhandle.

Comparing Machines

Wind machines are given a *rated output* by the manufacturer, which means simply the output of a particular model at a wind speed

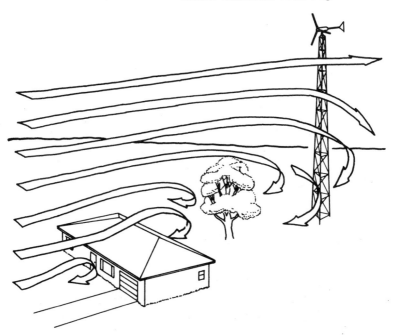

Figure 6–7:   The higher the tower, the fewer the obstructions to wind speeds. The distortion of wind currents occurs far above the height of the tree.

arbitrarily selected by the manufacturer. As you will see when you examine various models and their specifications, different models are rated at different speeds. How does one compare a wind machine that produces 1 k in 12-MPH winds with another that produces 2.5 k in 28-MPH winds?

Electrical engineer Gil Masters, one of the authors of *More Other Homes and Garbage,* provides a simple method by using the machine's capacity factor.[5] Remember that the capacity factor is the ratio of the energy delivered over a given period of time to the energy that would have been delivered if the generator were supplying rated power over the same time interval. Assume that a 10-kw machine delivers 1,200 kilowatt-hours (kwh) per month. If that machine were operating at full 10-kw-rated capacity for the 720 hours in a month, it would produce 7,200 kwh. Thus it has a capacity factor of 1,200 ÷ 7,200 or 16 percent. Figure 6–8, based on statistical work done by C. G. Justus in 1978, provides for correlations between wind speed, rated wind speeds and capacity factors. The capacity factor here is plotted against

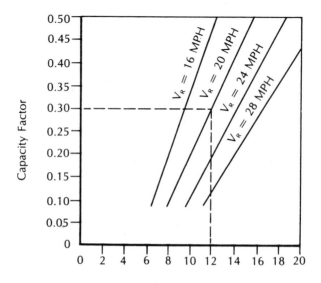

Average Wind Speed (MPH)

Figure 6 – 8:   Capacity factor is a function of average wind speed for different rated wind speeds (V_r). This graph is based on a regression analysis done by C. G. Justus. It assumes a cut-in wind speed of 0.4 V_r for all four of the rated wind speeds shown. The capacity factor is the fraction of the hours of the year the machine is operating at the equivalent of full output. To use this graph, you must know the rated wind speed of the wind machine you're considering and the average wind speed at your site. The dotted line shows the capacity factor that results from operating a machine with a V_r of 20 MPH at a site with 12 MPH average wind. Refer to the text to find out how to use capacity factor with a machine's rated power to derive an estimate of kilowatt-hour production. *From* More Other Homes and Garbage, *by Jim Leckie, Gil Masters, Harry Whitehouse, and Lily Young. Copyright © 1981 by Sierra Club Books. Reprinted by permission of Sierra Club Books.*

average wind speed with rated wind speed as a parameter. While the correlation assumes a cut-in speed of 0.4 V_r (V_r is *rated velocity*; for example, an 8-MPH cut-in for a machine with a rated velocity of 20 MPH), if a given machine has a different cut-in speed, it won't change the results significantly.

The example in figure 6–8 shows that a machine with a V_r of 20 MPH operated in a 12-MPH wind regime will have a capacity factor of 0.30. How does that translate into power production? Let's say the wind machine in question is a 4-kw unit. You can estimate monthly kilowatt-hour production by multiplying the capacity factor times the power rating times the number of hours in a month (720). Thus the equation becomes:

$$0.30 \times 4 \text{ kw} \times 720 = 864 \text{ kwh/month}$$

The difference between a machine's rated wind speed and your site's actual average is therefore very significant. If that 4-kw unit were rated at 24 MPH instead of 20 MPH, it would have a capacity factor of about 0.20, which, at a 12-MPH average, would deliver 576 kwh per month. That's quite a difference (33 percent) for just a 4-MPH change in rated wind speed.

James Sencenbaugh, a manufacturer of wind machines (see Appendix 6), adds another word of caution and some arithmetic advice in evaluating manufacturers' claims. He suggests that consumers compare the given machine and its claimed efficiency with those of known machines with long-term operating experience. To do this, one must use the Betz Theorem, which translated into a formula is: power in kw = K × A × V cubed, where K = 0.0000053; A = swept area of propeller = 3.14 × diameter (squared) ÷ 4; and V = rated wind speed of generator in MPH cubed.

The Sencenbaugh Electric Company provides a good example. They computed the average system efficiencies of several wind machines based on the rated wind speed, the rated capacity and the blade diameter. The results are listed below:

MODEL	EFFICIENCY (%)
Aerowatt 24FP7G	15
Dunlite 2 Kw	18
Jacobs 3 Kw 100 VDC	25
Pinson C2E	22
Sencenbaugh Model 500-14	24
Average of five machines	21

SOURCE: James Sencenbaugh. *Sencenbaugh Wind Electric Design Manual* (Mt. View, Calif.: Sencenbaugh Electric Co., 1981).

Assume a manufacturer claims 1.2 kw at 20 MPH, using a 10-foot-diameter blade. Using the formula above, one finds that the manufacturer is claiming to produce 3.3 k at 20 MPH. The claimed efficiency is thus 1.2 kw ÷ 3.3 kw = 36 percent. This manufacturer says he is 1.64 times more efficient than the average of the other six manufacturers. This comparison does not disprove the claim, but as Sencenbaugh cautions, it requires additional investigation.

Hydropower

A flowing stream contains two kinds of energy: Its velocity gives it *kinetic energy*, and its elevation gives it *potential energy*. The kinetic energy in most streams is not great enough to be useful. It is the potential energy between two sites of differing elevations that are exploited for hydropower. Very simply, the idea is to divert some of the water from a higher, upstream site, transport it via a conduit and then let it fall through a waterwheel or hydraulic turbine located at a lower elevation downstream. The turbine turns a generator that produces electricity, and the "spent" water returns to the stream.

The amount of power obtainable from a stream is proportional to the rate at which the water flows and the vertical distance the water drops (called the *head*). To determine the mechanical power generated, use the following formula: $THP = Q \times H \div 8.8$, where THP equals theoretical horsepower, Q is flow rate of water in cubic feet per second (CFS), H is head in feet and 8.8 is a correction factor for the units. For electrical power the proper formula is $P = QH \div 11.8$ or $P = AVH \div 11.8$, where P is the power obtained from the stream in

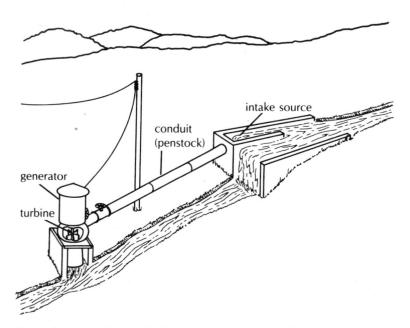

Figure 6–9: In this simple hydroelectric system, the flow of water turns the turbine inside the generator just as the power of the wind rotated the shaft inside the windcharger in figure 6–3.

kilowatts, Q is the flow of water in CFS; A is the average cross-sectional area of the stream in square feet; V is the average velocity of the stream in feet per second; H is the height the water falls (head) in feet; 11.8 is a constant that accounts for the density of water and the conversion from foot-pounds per second to kilowatts. Another formula, $P = (Q \times H) \div 709$, gives us the power in kilowatts with a known usable flow rate in cubic feet per minute (Q) and net head, in feet.

Using a flow rate of 10 CFM and a net head of 50 feet, the theoretical power is $P = (10 \times 50) \div 709 = 0.7$ kw.

These formulas give us an indication of the potential power from a stream of given dimensions and head. As in any conversion process, a great deal of potential energy is lost. Turbines themselves are from 60 to 85 percent efficient. Table 6–1 gives representative values for conversion efficiencies of various types of prime movers. Further losses take place in the generator and the gears. Thus the efficiency of the turbine should be multiplied by about 0.75 to give a system efficiency. A system efficiency of 50 percent represents a good preliminary estimate.

Any flowing stream can generate electricity. But below a certain combination of head and flow rate, the capital investment in the hydroelectric device will outweigh the benefits gained from electric gen-

---

TABLE 6–1
Typical Efficiency Ranges
for Small Water Wheels and Turbines

PRIME MOVER	EFFICIENCY RANGE (%)
Water Wheels	
Undershot	25–45
Breast	35–65
Poncelet	40–60
Overshot	60–75
Turbines	
Reaction	80
Impulse	80–85
Crossflow	60–80

NOTE: Reprinted and adapted from *Micro-Hydro Power*, Ron Alward, Sherry Eisenbart and John Volkman (Butte, Mont.: National Center for Appropriate Technology, 1979), p. 13.

Photo 6–5: One important part of a small hydro installation is the power-house, which houses the turbine, the generator and electrical and mechanical controls. This system produces 240 kwh per day during times when water flow is normal. During high-water times, it produces 480 kwh per day. *Photograph courtesy of T. L. Gettings.*

eration. Many developers believe that any combination of head and flow rate that produces less than 500 watts (0.5 kw), taking account of the system efficiency, is not worth the time and expense of development. Assuming a 50 percent conversion efficiency, this is about the power output from 450 gallons of water per minute (GPM) falling 9 feet. Or 20 GPM falling 200 feet. Or a flow rate of 1 cubic foot per second and a 12-foot drop, or a 6-foot drop (head) with a flow rate of 4 gallons per cubic foot. A flow rate of 1 cubic foot per second is equal to about 7 gallons per second.

On larger streams and rivers, one may be able to rely on published material to estimate heads and flow rates. The water flow is not constant even on rivers that have large reservoirs. It varies seasonally and in drought periods or periods of heavy rains and flood flows. In order to predict the power available from a particular hydro site, the historical water-flow records must be obtained and put in usable form. The majority of continuous flow data for a given river or stream are gathered and published by the U.S. Geological Survey (USGS). The most useful flow data readily available from the USGS is the average flow rate. Although hydropower equipment is usually rated at greater flow rates, the average flow rate can provide an adequate estimate for sizing the turbine on an initial, very preliminary basis.

In some cases, however, the USGS can provide, upon request, a flow duration curve for any given site. If the duration curve is available, the hydro plant will normally be sized for a 20 to 30 percent exceedance flow—that is, for the flow that is exceeded only 20 to 30 percent of the time. If a USGS gaging station is not located near the power plant site, reasonable approximations can be made using gaging station information, with appropriate adjustments, from a site within the same general rainfall area. The USGS will usually provide assistance in determining an approximate flow.

Measuring Water Flow

Assuming there is no reliable data for water flow that can be used on your site, and the state does not provide assistance, you will have to take your own measurements. Ron Alward, author of a manual on micro-hydro plants, recommends taking biweekly measurements throughout one complete year to get a good idea of how the flows vary seasonally. Then determine from available data for your watershed (such as precipitation or snow pack data) whether this has been a dry, a wet or an average year. Then make the necessary corrections to your own data to determine the minimum expected flow rates for your stream. This will give you the base capacity that can be counted on during the entire year. If, on the other hand, you are going to use the stream for shaving peak demand during certain months, or just want to maximize the revenue from the stream, you might want to size the system larger even though it will operate at lesser efficiencies during the periods of low flow. Some hydro developers have found it economical to run their systems even though a few months of the year the stream flow is so low that they have no output at all.

You will also need to know the maximum flow rate in order to size spillways adequately for bypassing excess water to prevent damage to your installation. There are three methods commonly used to determine flow rate. To measure the flow of small streams or springs, temporarily dam up the water. Divert the entire flow into a container of known size and carefully count the number of seconds it takes to fill the container. If it takes 40 seconds to fill a 55-gallon drum, the flow rate is 1.375 gallons per second or 82.5 gallons per minute or 11 cubic feet per minute.

For larger streams the *float method* can be used. If done carefully and repeated several times, it can give results accurate enough for most calculations. In order to use this method, you need to know the average cross-sectional area of the stream and the stream's velocity in feet per second. With this data you can find the flow, again in cubic feet per minute (area times velocity). These methods are detailed in the book *Micro-Hydro Power* (see the Bibliography).

A method for determining the flow rate of even larger streams is called the *Weir Method*. It is somewhat more complicated and more expensive than the other methods. You have to build a temporary structure across the stream perpendicular to the flow, with a rectangular notch or spillway of known proportions located in the center section. You then make certain measurements and use established tables to arrive at the volumetric flow rate (refer again to *Micro-Hydro Power*).

Measuring Head

Along with flow rate, determine the extent to which the water falls, which is where most of the power is created. The greater the vertical drop, the more potentially useful power there is available. In this chapter the general reference is to relatively low head (30-foot drop), as opposed to high head. To measure head you have basically three alternatives. You can hire a good surveyor to determine the vertical distance between your water source or proposed intake location and the proposed location of the power plant. Ron Alward suggests *in Micro-Hydro Power* that if your head is less than 25 feet, you may need such precise measurements that a surveyor should be hired. For those who know how to use standard surveying equipment (a transit or a surveyor's level and leveling rod), borrow or rent the appropriate pieces and get a friend to help you make the necessary measurements.

Once you have the total, or *gross head,* there are various losses to be considered before you can make any theoretical power calculations. The *net head* is required for these calculations. Losses will occur for several reasons. Whenever water flows through a pipe there are *friction losses*. These friction losses are greater for increased flow rates and for smaller pipe diameters. Elbows and bends in the pipe will also increase friction losses. Polyvinyl chloride (PVC) pipe, for example, offers low friction loss, rarely exceeding 8 percent of the gross head.

Other losses might occur that depend on the type of turbine you use, and there are power losses in the flow of water to and from the turbine. All these losses subtract fractionally from the *gross head* (what you measured) to give you the *net head*. When you know both the flow rate and the net head, you can make a determination of the available power in the stream (see figure 6–10).

Equipment

Most equipment suppliers can give you a rough estimate of the cost and type for your system if you can supply them with the following information: usable flow rate, length of pipe required from take-off to generator location (location of the dam with respect to the generator location), power demand (alternating current or direct current) and what you want to do with the surplus power.

There are two basic types of hydro turbines used to generate electricity: impulse turbines and reaction turbines. A third type, the Schneider Lift Translator, will be discussed because of its peculiar characteristics. Waterwheels will not be discussed here because their efficiencies are extremely low (about 15 to 25 percent), and it is unlikely that they would be the technology of choice for generating electricity. Each type and brand of turbine has its own variation in peak efficiency, efficiency at maximum load, efficiency at minimum load and ratio of maximum load to minimum load. Each of these variables affects the recoverable annual energy at a given site. Hence, knowledge of turbine types and manufacturers' data is necessary for the final turbine selection process.

Impulse Turbines

Impulse turbines are generally the simplest of all common turbine designs and are widely used in very small-scale hydro applications. Impulse turbines have a series of radial buckets that are driven by one or more water jets, each directed by a needle valve. Flow and power

(continued on page 212)

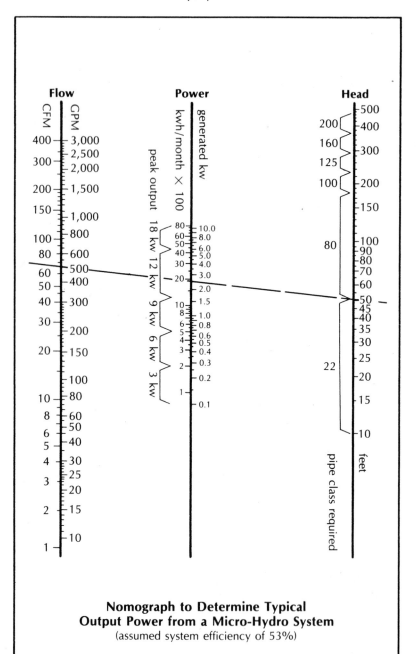

**Nomograph to Determine Typical
Output Power from a Micro-Hydro System**
(assumed system efficiency of 53%)

Figure 6–10:   Once you know the usable flow of your stream and the net head, you can use a nomograph (on facing page) to determine the power you can expect from your turbine. For example, suppose you have a usable water flow rate of 500 GPM through a net head of 50 feet. To determine the power you can expect from the turbine, locate the flow rate, 500 GPM on the Flow line, and the head, 50 feet, on the Head line. Join these two points with a straight line. The point where this line cuts the Power line is the power output of the turbine. The Power line gives three pieces of infor- mation. The continuous power output in generated kilowatts is 2.5. This turbine will put out a constant 2.5 kw if the water flow conditions remain the same. At a continuous power output of 2.5 kw, you can expect to pro- duce nearly 2,000 kwh per month, as indicated on the first scale on the left side of the Power line. If you are using a DC system, feeding into storage batteries, then your system can have a peak power output of 12 kw, as shown. *Redrawn from Alward, Eisenbart and Volkman,* Micro-Hydro Power, *p. 15*.

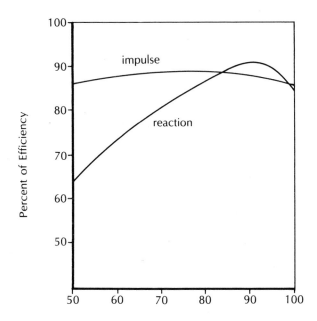

Turbine Loading, Percent of Rated Capacity

Figure 6–11:   This graph compares impulse and reaction turbine efficien- cies at different loads. Notice the relative stability of impulse turbines at partial loads, whereas reaction turbine efficiencies drop substantially at par- tial loads but are slightly higher than impulse turbines at 90 percent loads. *Redrawn from D. V. MacDonald and E. G. Raguse,* Technical and Eco- nomic Aspects of Pipeline Hydroelectric Projects *(Newport Beach, Calif.: Engineering Science), by permission of Engineering Science.*

are regulated by opening or closing the needle valve. The runner may be mounted with the shaft either vertical or horizontal and must rotate in air above the downstream water surface. Impulse turbines can maintain good efficiencies over a wide range of flows. These turbines are normally described by design head, power, speed and number of water jets.

Micro-Hydro Power succinctly describes the Pelton wheel: "In general terms a *Pelton wheel* is a disc with paddles or buckets attached to the outside edge. The water passes through a nozzle and strikes the paddles one at a time, causing the wheel to spin. The buckets are shaped so that the water stream is split in half and caused to change direction, heading back in the opposite direction to the original water stream for the greatest efficiency. The shape and smoothness of the buckets is important. Because the power developed by the Pelton wheel is largely dependent on the velocity of the water, it is well suited for high head/low flow installations."[6] Pelton wheels operate best on heads over 50 feet. Operating efficiencies of 80 percent are common.

Another impulse turbine is the *Turgo impulse turbine*, designed and built by F. Gilbert Gilkes and Gordon Ltd. of England. It is similar to the Pelton wheel but theoretically operates at a higher speed and thus can be used successfully at lower heads.

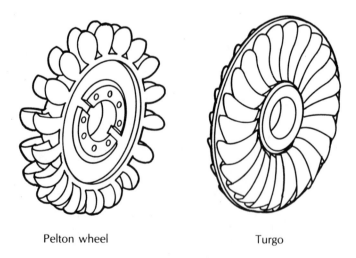

Pelton wheel   Turgo

Impulse Turbines

Figure 6–12

Reaction Turbines

Reaction turbines work differently from impulse turbines. The runner is placed directly in the water stream, and the power is developed by water flowing over the blades rather than striking each individually. Reaction turbines use pressure rather than velocity. The function is more like that of a centrifugal water pump running in its reverse mode. Reaction units tend to be very efficient at specific designed-for situations, and their efficiency falls sharply with any variation. Reaction units are usually used in very large installations.

The *Francis turbine* is a reaction turbine. According to Bill Johnson, editor of *Currents* magazine, it was the undisputed turbine of choice for nearly 100 years. Therefore, in the majority of cases where persons are rebuilding an existing turbine, they will be working with some variation of the Francis turbine. By the same token, nearly every used unit offered for sale will be a Francis. Francis turbines must be mounted in a fashion that allows the water to enter its rotors from all sides. It can be mounted horizontally or vertically. It has been used for heads from 3 to 300 feet.

The *propeller turbine* is a simple machine, closely resembling a boat propeller in a length of pipe. As with the Francis, the water

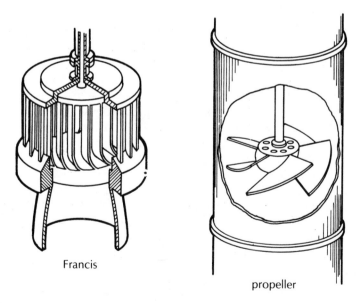

Francis

propeller

Reaction Turbines

Figure 6–13

contacts all of the blades constantly, and it is thus imperative that the pressure in the cross section of the pipe be uniform. If, for example, the unit were operating horizontally, and the pressure at the top of the tube were less than at the bottom, the runner would be out of balance. The propeller turbine is an axial-flow reaction turbine in which the water moves parallel to the runner axle and turns the blades as it passes. Large versions of this are sold by Allis Chalmers under the trade name TUBE Turbine. The *tube turbine* bends the penstock just before or after the runner, allowing a straight-line connection with the generator, which is located outside the pipe. Another variation is the *bulb turbine*. In this case, the propeller drives a gearbox and generator in a pod or "bulb" within the tube holding the propeller. This makes for a very compact, self-cooled unit without the problems inherent in other types of sealing the axle where it leaves the tube. However, it also requires a high degree of precision to make the gears and generators as small as possible and involves some difficulty in servicing because of the lack of easy access. The bulb turbine places the turbine and generator in a sealed unit directly in the water stream. Another variation, the *straflow*, attaches the generator directly to the perimeter of the turbine. The *Kaplan* has adjustable blades on the propeller to allow for variations in flow rates.

Cross-Flow Turbine

The *cross-flow* turbine is midway in style between an impulse and a reaction turbine. Manufactured primarily by the Ossberger organization of West Germany, the cross-flow turbine has a horizontal axle with an attached drum composed of curved blades parallel to the axle. A nozzle directs the water stream into the blades from the outside, after which the water passes through the drum and out via the blades opposite where it entered. It was apparently invented as an impulse wheel similar to the Pelton wheel but has been refined to have a reaction component by using a two-element nozzle. The cross-flow turbine is not capable of the efficiencies of a well-tuned Francis or variable pitch propeller turbine, but has the potential of maintaining efficiency over a wide range of flows in its more sophisticated forms. It also has the advantages of two axle ends on which to attach generators, pumps, and so forth and has no thrust bearings or cavitation to contend with. The Ossberger cross-flow turbine can work effectively on heads of less than 20 feet.

Several companies have been examining the feasibility of using pumps as turbines. Pumps offer the advantage of being off-the-shelf items, while most turbines must be custom designed. They cost about half the price of a turbine of equal capacity. On the other hand, they

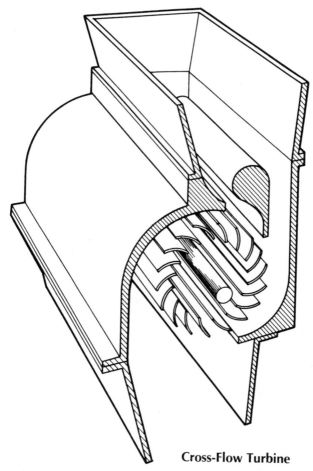

Cross-Flow Turbine

Figure 6–14

are about 5 percent less efficient than turbines, and they are less capable than turbines of operating over a variety of flows. Therefore, several pumps might be necessary to satisfy the flow requirements of a particular site. Costs should be less by about 50 percent. *Centrifugal Pumps as Hydraulic Turbines for the Small Hydropower Market,* by L. Shafer and A. Agostinelli, is an excellent publication on the function of centrifugal pumps as water turbines (see the Bibliography).

Schneider Hydroengine

A recent development in hydro-turbine technology is the Schneider Lift Translator or Hydroengine. Daniel J. Schneider, a

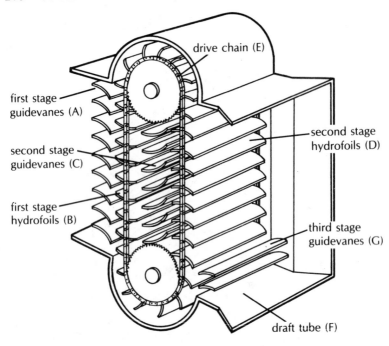

first stage guidevanes (A)

drive chain (E)

second stage guidevanes (C)

second stage hydrofoils (D)

first stage hydrofoils (B)

third stage guidevanes (G)

draft tube (F)

Schneider Hydroengine

Figure 6–15: *Adapted by permission of Schneider Engine Co. from Daniel J. Schneider and Emory K. Damstrom,* Schneider Hydroengine.

physician-turned-physicist, was seeking a very low head hydropower plant that could harness the vast majority of water flowing in the United States without high dams that flood thousands of acres. Picking up on the work of another physician/physicist, the eighteenth-century Swiss Bernoulli, he examined the principles of fluid flow dynamics. Schneider used the principles of flow dynamics that apply to wind machines. Indeed, he first designed his machine to be used as a wind turbine. Historically, lift translators have been used for sailboats, airplane wings and hydroplane foils.

Figure 6–15 illustrates the operation of the patented Schneider Hydroengine. The energy converting process begins as the water flows into the entryway. The design of the entryway causes the water to strike the cross section of the Schneider Hydroengine at a uniform velocity. "As the water flows into the throat section, it contacts the first stage guidevanes (A). The first stage guidevanes direct the water flow to lift downward on the first stage hydrofoils (B) by directing the flow to optimally match the velocity of the hydrofoils. This velocity matching achieves a high recovery efficiency. Once the water flow leaves the first stage hydrofoils, it comes in contact with the second stage guide-vanes (C). The second stage guidevanes redirect the water velocity in an upward direction, lifting the second stage hydrofoils (D), at a matched velocity. The hydrofoils' linear movements are transferred to drive chains (E) attached to each end of the hydrofoils. The drive chains turn sprockets located at the top and bottom of the engine. The top sprockets are attached to a shaft leading to the speed increaser [transmission gearbox], and the shaft's torque/rotation drives the generator. After leaving the second stage hydrofoils, the water flow simultaneously enters the draft tube (F) and the third stage guidevanes (G) for its return to the channel. The third stage guidevanes and draft tube are designed to minimize head losses. The draft tube is horizontally flared and the outlet is at an elevation so that the top margin is never less than six inches below the lowest operating water level in the afterbay."[7]

This machine will economically derive 80 percent of the power output from a given flow of water from a head substantially less than that needed by a conventional machine. It can operate well in heads of only 3 feet and is presently available in sizes ranging from under 1 kw to 1 megawatt. The Schneider Hydroengine is available from the company in Justin, Texas (see Appendix 6).

Governors, Generators, and Powerhouses

With today's technologies and the ready availability of utility interconnection, the conventional mechanical or hydraulic *governor*, a device to change the speed of the machine, is seldom necessary for systems below 200 kw. If an induction generator is used, no governor is necessary. If you are considering an isolated system, one not tied to a utility grid, several types of electronic governors are available. Most of these use the surplus energy to generate heat, often for hot water heating, in effect generating at a constant rate and shedding excess as heat.

Generator choices in the 1-kw to 5-kw range are limited. The ultimate choice will depend to some extent on the RPM of your turbine. Generators generally run at 900 RPM, 1,200 RPM or 1,800 RPM, with 1,800-RPM being by far the most common and the cheapest. Direct-drive generators are normally only used for either very high-speed installations such as high-head Pelton wheels or very large low-head turbines. Otherwise geared transmissions are used to step up shaft speed to the generator.

J. George Butler, author of *How To Build and Operate Your Own Small Hydroelectric Plant*, describes several key factors for deciding on the location of the powerhouse. First, it should be located so as to obtain the greatest possible amount of head. Second, it should be as close as possible to the house, given the winter snows that would make walking to the powerhouse difficult. Third, it should be located high enough off the stream so as not to be damaged by spring runoffs. The powerhouse must be large enough for the turbine, generator, main breaker box, inverter and battery storage, if any. Its design need not be sophisticated, but it must be weatherproofed and have a dry floor. The cost of the completed structure should be in the range of $10 to $20 per square foot, depending on how much salvaged materials were used and how much of the work is done by the owner.[8]

Photovoltaics

The most decentralized of all electric technologies is the photovoltaic (PV) cell, more popularly known as the solar cell. Unlike the other three technologies discussed in this book, the solar cell doesn't convert mechanical energy into electrical energy.

The vast majority of solar cells today are made from silicon. The silicon is reduced from silicon dioxide, ordinary sand. The silicon used in solar cells is very pure, on the order of one atom per billion impurity. The silicon atom is called a *semiconductor* because it has four electrons in its outermost shell. Those elements with only one or two electrons in their outer shell have weak holds on the electrons. These tend to move easily from one atom to another. Elements composed of such atoms are called metallic. They readily conduct electricity—which is itself nothing more than a flow of these outer shell electrons. If an element has as many as six or seven electrons in its outer shell, the material is called nonmetallic. The electrons are tightly bound, and the material is a poor conductor of electricity.

Elements whose atoms have three, four or five electrons in the outer shell, are called semiconductors because their outer electrons can be freed, but only if given some additional energy, or push, from an outside force or energy. Silicon is a semiconductor.

Sunlight is composed of tiny bits of energy called *photons*. A silicon atom readily absorbs a photon, and the added energy "excites" or activates one of the outer electrons and frees it. (This occurs most easily when silicon atoms are lined up in precise rows or positions, which they are in their crystalline state. In any crystal the atoms or molecules are arranged in perfect geometric formations. The opposite is the *amorphous state*, in which atoms or molecules are jumbled together in no regular pattern.)

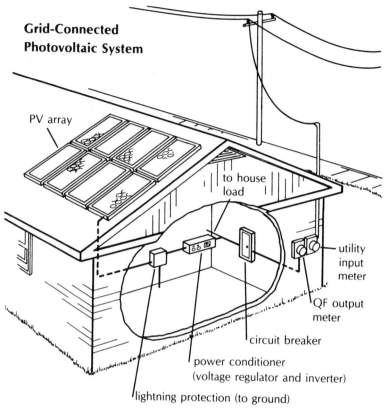

Grid-Connected Photovoltaic System

PV array

to house load

utility input meter

QF output meter

circuit breaker

power conditioner (voltage regulator and inverter)

lightning protection (to ground)

Figure 6–16

The silicon atom loses its outer shell electrons when struck by sunlight, but this alone does not generate electricity. To do that, the flow of these electrons must be channeled. In the cell, a junction or barrier is formed within the wafer of silicon so that when the photon moves the electron out of orbit, it moves into a wire that is part of a circuit, at which point it creates with other electrons a channeled current of electric power.

This barrier is created by adding infinitesimal amounts (no more than one atom per million atoms of silicon) of two elements: boron, which has three outer shell electrons, and phosphorus, which has five. The two surfaces of the cells are treated with impurities, creating dissimilar electrical properties from side to side. In the process, the cells are arranged on a rack and inserted into a heated chamber. One side of the cell is exposed to positively charged boron atoms and the other

Photo 6–6: Future I, a solar research house sponsored by Georgia Power Company, uses 544 square feet of photovoltaic panels to provide electricity for the house. Future I is connected to the Georgia Power Company utility system so that at night and on cloudy days electric service is available. When the photovoltaic cells are producing more electricity than the house needs, the homeowner sells the excess to the utility. *Photograph courtesy of Georgia Power Company.*

is exposed to negatively charged phosphorus or arsenic atoms. This negative charge forms "holes," as they are called, to attract electrons from the boron atom. As the wafer is heated, these elements are allowed to penetrate the surface of the cell. The amount of penetration is a highly controlled process. As the two sides of the cell are exposed to the desired depth with these different materials, a neutral area is left between the sides. This neutral area is called a *PN junction*. It gets its name from the positive and negative sides that it separates. The side of the cell treated or doped with boron needs electrons and tends to absorb them, much as a sponge soaks up water. The side treated with phosphorus atoms has an excess of electrons, which creates a negative charge.

When light falls on the cell, photons are absorbed and electrons are set free. The surplus electrons accumulate in the phosphorus-treated side. If one end of a wire is attached to this top layer and the other end connected to the layer beneath, electrons will leave the

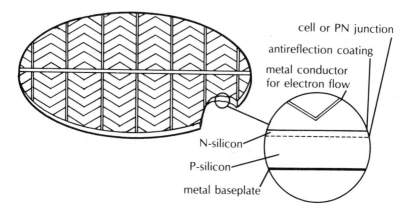

Crystal Silicon Solar Cell

Figure 6–17

upper layer, flow through the wire, and be absorbed by the boron-doped silicon, completing the circuit.

The final production step for the solar cell consists of an etching process that applies a metal coating. A metallized grid is applied to the surface of the cell to gather up the electrical charges. The lines of this miniature grid are extremely thin so as not to shade the cell from sunlight. On the other hand, the grid must cover all parts of the cell. If its lines were spaced too far apart, the electron might not have the energy to travel all the way from its orbit to the grid. The grid appears like a system of roads all leading to a highway.

Finally, the top side of the cell containing the grid is covered with a special antireflective coating to prevent sunlight from being reflected off the cell. This increases the amount of sunlight absorbed and improves the overall efficiency of the cell. This antireflective material also has a textured surface, which further increases the amount of light absorbed by the cell.

Yet despite all technological refinements and extreme quality controls, solar cells typically convert a fraction of the available solar energy into electricity. The maximum theoretical efficiency of the silicon solar cell is 23 percent. Efficiencies of 20 percent have been achieved in the laboratory, though commercial, single-crystal silicon solar cells typically convert 13 to 15 percent of the sunlight falling on them into electricity.

The finished cells are then packed into modules, which consist of several cells connected in series until the desired voltage or amperage

solar cells in series (increases voltage)

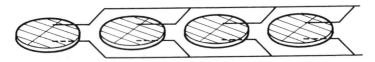

solar cells in parallel (increases current)

Figure 6–18:   When wired in series, each solar cell voltage output is added together. When wired in parallel, the currents are added together.

is achieved. These cells are then housed in a protective covering, usually glass on top with silicon rubber holding the cells in suspension. A rigid frame is placed around the glass to protect it as well as to provide some means of mounting.

Electrical Characteristics

The amount of current (amperage) produced by a photovoltaic cell is proportional to the amount of light falling on the cell. Current also increases with the area of the cell. The voltage, on the other hand, is fairly constant, about 0.45 v in silicon cells regardless of cell area. Unlike conventional generators, the cell current decreases with increases in cell voltage. Also, the output of a silicon-based cell is reduced by about 10 percent for each 45°F rise in cell temperature.

Based on the application and load requirements, the main terminals of a solar cell array are connected to various power conditioning equipment. The direct current produced by the array can be used directly or converted to standard 120-v alternating current.

Because the array is an active power source by day, care must be taken during servicing and maintenance. The National Electrical Code recommends that in the interest of safety, maximum voltage should be no higher than 50 v to 60 v. If a higher voltage is required, switches should be installed between cell modules at 50-v to 60-v intervals. These switches can then be disengaged during maintenance to limit the voltage anywhere on the array.

Cost of Solar Cells

Solar cells have been included among the four technologies discussed here even though they are not yet economical for most grid-connected applications. They are, however, already competitive for many stand-alone applications. When a house is located far from the grid system, the cost of laying in cable may be more than the cost of a PV system. More than 1,000 homes in the United States had solar cells as of the end of 1982. And although still in the demonstration stage, at least a dozen grid-connected, solar-cell-powered homes are operating.

Before 1973 solar cells were used only in the space program. They were manufactured for the highest possible reliability. Their first market

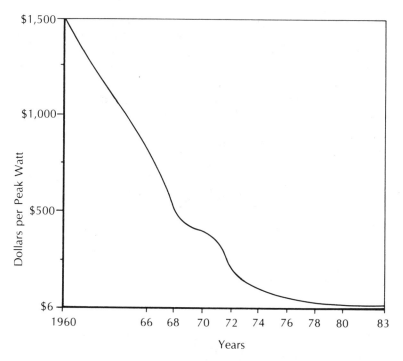

Figure 6–19: This graph shows how the prices of photovoltaics have fallen over the years. The prices indicated here are prices per peak watt of capacity for large quantities of photovoltaic modules. Prices are factory costs, not retail or wholesale prices. For a large quantity of photovoltaics, such as 1 megawatt's worth, which equals $10 million in hardware, the price can be as low as $6 per peak watt.

on earth was for extremely remote applications, at the North and South Poles or in deserts or on mountaintops where their high cost was less than that of heavy-duty batteries or diesel generators that had to be serviced regularly. But the more the cost of cells drops, the more they will become an increasingly popular consumer item.

The price is in fact dropping quickly. In 1974 a photovoltaic system big enough to make a household energy self-sufficient would have cost $2.5 million. Today it costs $50,000, a fiftyfold reduction in less than ten years. Some members of industry and government predict that solar cell electricity will be competitive for grid-connected applications by 1985 or 1986, assuming a continuation of the business and residential tax credits. At that time a household system would cost between $10,000 and $15,000, and would be economical as a grid-connected system.

Another type of cell, the *concentrator*, is nearly cost-competitive right now. These use lenses or mirrors to focus the sunlight on very high efficiency cells made of single crystal silicon or gallium arsenide. The concentrators are made of plastic, glass or aluminum. These materials are cheaper than semiconductor-grade silicon, and as concentrators they cost-effectively reduce the required cell area. Unfortunately, concentrating sunlight on solar cells increases their temperature and thus decreases their efficiency if the cells are formally cooled. But efficiencies do not drop at higher temperatures for cells made out of materials such as gallium arsenide. Several companies are also investigating the possibility of solar cogeneration systems that generate heat as well as electricity. The heat gained from cooling the cells (to maintain efficiency) could be used in industrial processes.

In fact, even the flat cell models that we used on rooftops may be somewhat more cost-competitive when thermal as well as electrical energy is captured. By using air or water to cool the cells and using the resulting heat for space and water heating, solar cogeneration can raise overall efficiencies from 15 percent to more than 50 percent, with a theoretical upper limit of 70 percent.

Finally, scientists have discovered a way to achieve relatively high efficiencies even with very "dirty" silicon. Some manufacturers have produced thin film solar cells from amorphous silicon with efficiencies of 8 percent. Since a very small amount of material is used, and the crystallization step is avoided, amorphous silicon cells are very inexpensive. Since 1981 the Japanese electronics industry has routinely used amorphous silicon in watches and calculators. Solar arrays using such devices will generate less electricity per square foot, but they might become competitive as early as 1984.

Measuring the Sun's Energy

Manufacturers sell photovoltaic systems on the basis of the cost per peak watt of the cell. A *peak watt* is the electricity generated from a solar cell under peak solar gain conditions. But the sun doesn't shine all day and neither does it shine at full capacity even during daylight hours. A typical clear day will contain 4 to 6 hours of full sunlight equivalent. A very rough rule of thumb is that to get 1 average watt of power, one needs 4 to 6 peak watts worth of installed cells. An average watt is the amount of power one cell would generate if operating continuously over the year. Thus to generate an average of 720,000 watt hours (WH) or 720 kwh per month, 1,000 watts (w) per hour times 24 hours times 30 days, a system would need 4,000 to 6,000 peak watts of installed capacity.

This power requirement ultimately translates into some number of square feet of cells covering a roof or arrayed on the ground. But that number is highly variable depending on cell efficiency, cell shape and the level of solar gain at a given site. Another rough rule of thumb has that in the United States 1 square foot of cells will deliver 7 to 10 peak watts.

The power generated by the solar cell array varies depending on the tilt of the array. If the tilt angle is equal to the latitude of the installation, the noon insolation will be the same in winter as in summer. However, the array will produce a great deal of electricity in the summer due to the longer summer days, and a small amount of electricity in the winter. By varying the tilt of the array one can even out the differences between winter and summer sun angles. The extent to which the solar cell owner will do this depends on the buyback rates of the local utility under PURPA and whether there is a seasonal and time-of-day component to them. For example, a more horizontal tilt will gather more energy during the summer months, but little insolation during the winter when the sun lies low in the sky. But daylight hours are longer during the summer, and if the summer rates are higher than winter rates (probable in most of the country, which is served by summer peaking utilities) the solar cell owner could have a great incentive to lower the tilt of the array.

The optimum tilt of the array for year-round generation is usually given as 15 degrees plus latitude. Thus for Memphis, Tennessee, it would be 35 degrees latitude plus 15, or a 50-degree tilt.

Table 6–2 shows the electricity in kilowatt-hours (rounded off to the nearest watt hour) generated per square foot per month in Memphis for two array tilts. Option 1 has a tilt of latitude plus 15 degrees. In

TABLE 6–2

Kilowatt-Hours of Electricity Generated
per Square Foot per Month (10% efficiency)

	JAN.	FEB.	MAR.	APR.	MAY	JUNE	JULY	AUG.	SEPT.	OCT.	NOV.	DEC.
Option 1	1.0	1.1	1.3	1.4	1.5	1.4	1.5	1.5	1.4	1.3	1.0	0.8
Option 2	0.8	1.0	1.3	1.6	1.8	1.8	1.9	1.8	1.5	1.3	0.9	0.7

NOTE: Option 1 has an array tilt of latitude plus 15 degrees; option 2 of latitude minus 15 degrees.

Option 2 the array tilt is latitude minus 15 degrees.

The first option gives a more balanced generation pattern. The ratio of lowest to highest month is less than 2:1. The second option generates more electricity but is more unbalanced. The ratio of lowest to highest month is almost 3:1.

A 1,000-square-foot array under the first option would generate 15.2 kwh × 1,000 = 15.2 megawatt-hours (Mwh) per year. Under the second option, a 1,000-square-foot array would generate 16.3 kwh × 1,000 = 16.3 Mwh per year, or 7 percent more. Assuming a seasonal or time-of-day rate differential for a summer peaking utility, the second option is even more favorable, generating 20 to 25 percent more electricity in the May-through-August peak air-conditioning load months.

Cogeneration

Cogeneration systems produce both electrical (or mechanical) energy and thermal energy from the same primary energy source. The automobile engine is a good example of a cogeneration system. It provides mechanical shaft power to move the car, produces electric power with the alternator to run the electrical system and uses the engine's otherwise wasted heat to warm the inside of the car in the winter. Cogeneration systems capture otherwise wasted thermal energy, usually from a heat engine producing electric power (i.e., a steam or combustion turbine or diesel engine), and use it for applications such as space conditioning, industrial process needs, or water heating, or as an energy source for another system component.

Among the four technologies discussed here, cogeneration is unique because it is not necessarily based on renewable fuels

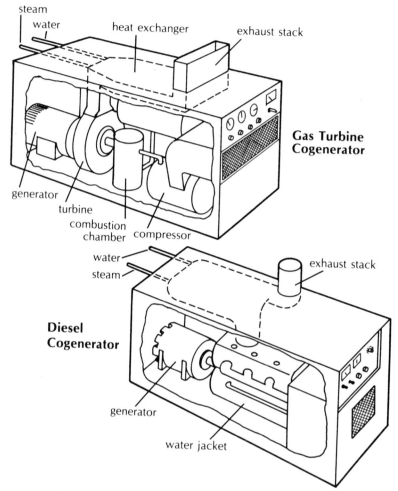

steam
water
heat exchanger
exhaust stack

Gas Turbine Cogenerator

generator
turbine
combustion chamber
compressor

water
steam
exhaust stack

Diesel Cogenerator

generator

water jacket

Figure 6–20: Shown here are the two basic types of cogeneration systems: the gas turbine and the diesel engine.

(although it can be if fueled by wood, solid waste or methane). In fact a political battle took place over whether to include it at all in PURPA. Some utilities argued that encouraging gas or oil-fired cogeneration could actually increase the amount of domestic and imported fossil fuels burned to generate electricity, a result clearly antithetical to the objectives of the legislation. Some argued that many small cogeneration facilities that burned oil could represent a potentially greater pollution agent than a few large, oil-fired power plants because the latter have taller smokestacks, are located far from population areas and have more sophisticated antipollution devices.

The Federal Energy Regulatory Commission (FERC) decided to include oil- and gas-fired cogenerators under PURPA but only after delaying for more than a year before concluding that the environmental impact would be benign. The FERC argued that even if slightly more fossil fuels were used for a cogenerator than for either a boiler or a conventional power plant, the amount of fossil fuel used would be less than the *combined* amount of fuel burned to generate the same amount of electricity and thermal energy in both the power plant and the boiler. A typical power plant operates at 30 to 35 percent efficiency. A typical boiler operates at 50 percent efficiency. Cogeneration systems operate at 70 to 95 percent efficiencies. For every 100 barrels of oil burned in a cogeneration plant, about 10 percent less electricity would need to be generated than in a conventional power plant and in addition, the cogeneration system would supply almost as much heat as would be generated in a conventional boiler.

There are two types of cogeneration systems: topping cycle and bottoming cycle. A *bottoming-cycle* system uses reject steam to generate electricity. The *topping-cycle* system generates electricity as the primary product and recovers the waste heat. This chapter discusses only topping-cycle cogeneration systems, since only these are available with outputs of less than 200 kw. Commercially available cogeneration systems within this size range are diesel or gas turbines or reciprocating piston engines.

Cogeneration systems consist of three primary components: the engine, the generator and the heat recovery equipment. Very few companies sell a complete, packaged system. Almost all companies combine the best individual components into the most cost-effective system.

The gas in gas turbines does not refer to the fuel, but rather to the gases that provide the force to spin the turbine. The fuel may actually be oil or natural gas. In a simple gas turbine, the air is compressed and then pushed into a combustion chamber. Fuel then enters the chamber and burns, raising the temperature of the gases. As the gas expands and escapes through the nozzle, it rotates a turbine. The spinning shaft turns a generator, and also provides the power for compressing additional air. In gas turbine cogeneration, the hot exhaust gas can heat water in a boiler to produce steam. Gas turbines have a lower electrical efficiency than diesel engines (20 to 30 percent) but the total recoverable energy is about the same. The electrical-to-thermal ratio for simple gas turbines is between 0.5 and 0.7 (50 to 70 percent of the output is electric power). This means that as much as twice as many Btu of heat energy are recovered for every Btu of electrical energy generated. The

Photo 6–7:   A small-scale cogeneration system such as this one can operate independently to supply the energy needs of an office building, an apartment building or several homes, or it can be grid connected. In the latter case, any surplus electricity produced can be sold to the local utility under PURPA regulations. *Photograph courtesy of Agway Research Center.*

waste heat in gas turbines can be recovered at a higher temperature, in the range of 900°F to 1,000°F.

In sizes much smaller than 1,000 kw, the fraction of primary energy converted to electricity is very low, often less than 20 percent, compared to 30 percent for larger systems. Also, the electrical conversion efficiency of gas turbines drops significantly in part-load applications. Thus an oversize system will operate much more inefficiently for a gas turbine than would be the case for a diesel-based system. The result is that for the size of systems discussed in this book, gas turbines are usually uneconomical. However, advances in the technologies could change this conclusion.

The high-temperature gas turbine exhaust heat can be used to increase power generation. Turbine exhaust temperatures would be reduced several hundred degrees for this purpose and would still be hot enough to meet most domestic and commercial thermal requirements. One way this can be done is to run turbine exhaust through a regenerator, which transfers heat to the air coming out of the compressor before it goes to the combustor. General Electric predicts that the electrical efficiency of cogenerating regenerative gas turbines could reach 38 percent by the mid-1980s. These systems produce lower-temperature heat and also less net heat (in Btu) than simple gas turbines, so the percentage of primary fuel converted into useful thermal and electrical energy declines. But since the proportion of electrical energy goes up, the economic value of the output rises. The electric output of cogeneration systems is the real money-maker. At 5¢ a kilowatt-hour, the value of the electricity is about $15 per million Btu. The value of thermal energy at a fuel oil cost of 90¢ per gallon or a natural gas price of 60¢ per therm is about $7.50 and $6.00 per million Btu respectively.

The gas turbine can be modified to operate on methane as long as this low-Btu gas (usually 600 Btu per cubic foot compared to 1,000 for natural gas) is clean enough for gas turbine operation. The primary modification entails increasing the fuel gas flow rate fivefold to tenfold to make up for the lower heating value of the fuel compared to natural gas. Gas turbines operating on low-Btu gas can actually be 10 to 20 percent more energy-efficient than when operating on natural gas. Moreover, a given gas turbine would also have a higher power output when operated on low-Btu gas. It is also possible to modify gas turbines for dual fuel operation so that a low-Btu, gas-burning gas turbine might be switched to oil or natural gas, depending on fuel price differences and availability.

The other basic type of cogeneration system involves diesel engines driving a reciprocating piston. Once again the name is misleading. Diesel engines, which can be as small as 15 kw, can be fired by natural gas or oil. Diesel engines can either be spark-ignited like an automobile gasoline engine or compression ignited. Spark-ignited stationary engines are usually fueled with natural gas, while compression ignition engines are usually fueled with diesel fuel or residual fuel oil, although there are a number of dual fuel engines available. When switched to low-Btu gas, the diesel will produce less power and may be less efficient in its electrical conversion efficiency, in contrast to the situation with gas turbines.

Diesels convert a higher fraction of the primary fuel to electricity than do gas turbines, about 35 percent versus 30 percent. Diesel cogeneration provides by-product heat at both high and low temperatures. Diesels use a water jacket to cool the cylinders in which water is heated to 150°F to 250°F. Exhaust gases are emitted at about 700°F. Even though less waste heat is usually recoverable with these engines than with gas turbines, the higher electrical conversion efficiency makes their total output (thermal and electrical) more economically valuable.

There are no household cogeneration systems yet available. The smallest one is the Total Energy Module System (TOTEM) distributed by the Fiat motor company. Yet it has been used primarily in Europe as a

Electricity Only

Process Heat Only

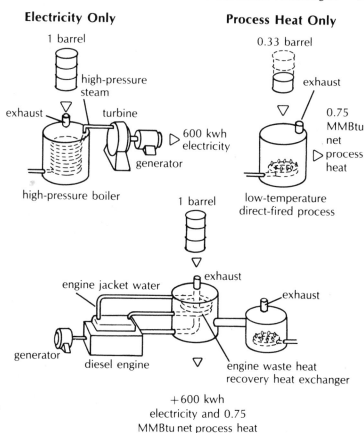

1 barrel

high-pressure
steam

exhaust

turbine

600 kwh
electricity

generator

high-pressure boiler

0.33 barrel

exhaust

0.75
MMBtu
net
process
heat

low-temperature
direct-fired process

1 barrel

exhaust

engine jacket water

exhaust

generator

diesel engine

engine waste heat
recovery heat exchanger

+600 kwh
electricity and 0.75
MMBtu net process heat

Electricity and Process Heat

Figure 6–21: This illustration shows that diesel cogeneration can result in significant oil savings. The figures at the top (electricity only and process heat only) show that the production of 600 kwh of electricity and 0.75 million Btu of low-temperature process heat consume a total of 1.33 barrels of oil. The lower figure (electricity and process heat) shows that the same output results when the two processes are combined, but the consumption is 1 barrel of oil. The use of waste energy results in saving one-third of a barrel of oil. *Adapted from Governor's Commission on Cogeneration*, Cogeneration: Its Benefits to New England *(Boston: Massachusetts Energy Office, 1978).*

peak-shaving device. It is turned on rarely to meet the peak needs, thereby reducing the heavy demand charges utilities impose year-round for customers that have sharp demand spikes for only a few minutes a year.

If the TOTEM were used as a baseload facility operating 90 to 100 percent of the time, it would generate at rated capacity about 130,000 kwh per year and almost twice that amount in useful thermal energy. This could meet the needs of almost a dozen homes. Thus if homeowners are to gain the advantages of cogeneration, they should form collectives, possibly homeowner associations, to own and operate the system, much the way they now own and operate swimming pools or other recreational facilities.

The problem with operating the TOTEM system as a baseload technology is its high maintenance cost. Automobile engines are not made to operate under a high, constant torque. The 450-horsepower (HP) engine under the hood of the Cadillac may need a great deal of power to get the car going. But on the highway, cruising at 60 MPH, it probably needs only 15 HP to move the vehicle. Automobile engines, unlike truck engines, are not made to operate under high torque on a sustained basis. Thus the TOTEM system must be completely torn down and maintenanced every six months at a cost of several thousand dollars.

About a dozen TOTEM systems are operating currently in the United States. Several are on farms, where they operate from methane generated from animal wastes. If the system can overcome its high maintenance costs, it has several attractions. It is a complete system. All electrical controls and heat recovery equipment come in one package. Because of this, the space required for that equipment is much smaller than for a system assembled from parts from different manufacturers.

Two companies at present sell and install cogeneration systems in the 75-kw to 200-kw range: Cogenic Energy Systems, of New York, and Re-Energy Systems, Inc., of Media, Pennsylvania. Re-Energy Systems is discussed in more detail here, primarily due to its claimed higher efficiency and the five-year warranty the company provides. Moreover, Re-Energy Systems is more oriented to smaller commercial users than Cogenic Energy Systems.

Re-Energy Systems uses a Mack truck engine and operates it at about half the rated horsepower to give it a longer life. The engine runs at about 900 RPM to 1,200 RPM compared to TOTEM's 3,000 RPM.

The generator is made by a local firm in Pennsylvania and matched to the engine. A 100-HP engine will run a 75-kw generator.

Three heat exchangers are mounted in a heavily insulated, all-steel enclosure. One exchanger reclaims the heat from the lubricating oil, heated air, the generator's resistance losses and heat radiated from the engine block. Another takes heat from the cooling jacket of the engine. The third absorbs sensible heat from the exhaust system and from the condensation process as well. One reason Re-Energy can

claim an efficiency of 96 percent compared with an average 80 percent
· for their competitors is that its units capture the waste heat from exhaust
gases. The temperature coming out of the engine is 1,200°F while the
temperature of the fluid finally dumped is 90°F. The unit achieves
96 percent thermal recovery from the generation of electricity. The
company claims the unit when fully loaded converts 38 percent of the
primary fuel to electricity and 58 percent of the primary fuel to recovered thermal energy.

Depending on the type of cogeneration equipment used, one can
maximize heat recovery or electrical generation or maximize the recovery of high-temperature heat. Most cogeneration systems produce
twice as much heat as electricity.

Until PURPA, a cogeneration system was invariably sized to meet
the internal base electrical load of the building, unless it was used for
emergency backup. (All hospitals, for example, are required to have
standby generators to power life-support systems in case the utility
lines go dead.) Sizing a system for the internal base electrical load
allows the owner to use all of the electricity internally and to dump a
small amount of the waste heat during summer months. Often the
cogenerator was used as a peak-shaver, turned on during peak times of
the year to reduce expensive demand charges.

PURPA's more favorable standby and buyback rates encourage
larger cogeneration systems because surplus electricity can be sold to
the utility. At an average buyback rate of 5¢ per kilowatt-hour, the
value of the electricity is $15.00 per million Btu (MMBtu). At a price of
$1.25 per gallon of fuel oil, the value of the heat energy generated
(assuming one can use all of it on-site) is $10.00 per MMBtu. When
buyback rates are high in proportion to the cost of the fuel for the
cogenerator, the incentive is to maximize the electrical output. On the
other hand, if the buyback rate is derived from a coal-fired utility with
no new power plants under construction, the buyback rate might
be only 2¢ per kilowatt-hour, for a value of $6.00 per MMBtu for the
electricity. If fuel oil is priced at $1.25 per gallon, the value of usable
thermal energy is still $10.00 per MMBtu. In that case the cogenerator
would be sized only to meet the internal electrical base load.

Assuming high buyback rates, the trade-off then is between the
higher cost of a larger system and the higher amounts of waste heat
generated during the nonwinter months, unless that heat can somehow
be used. The advantage of a cogenerator is only fully captured when
the full thermal as well as electrical output is used. At total thermal plus
electrical efficiencies of 80 percent, a cogenerator will be burning
slightly more fuel to produce the same amount of heat plus a great deal
of valuable electricity. The electricity is thus "free" (discounting the

higher cost of the cogeneration equipment than the conventional boiler) but it is not free if the thermal energy is not used. Indeed, because the electrical efficiency of most cogenerators is the same or less than that of peaking power plants and their cost per kilowatt of capacity is greater, the cost of electricity generated in this fashion will be higher than the marginal cost of electricity to the utility.

The conclusion is that on-site power plants cannot compete with central power plants without recovering a significant percentage of their waste heat. The only time that this might not be true is where the central power plant burns expensive fuel, such as oil, and the on-site power plant uses methane, waste wood or some other local fuel source at relatively little collection expense.

Thus, when sizing a system, there is a trade-off between the amount of electricity one can export to the grid system and the amount of heat that can be usefully captured internally. As a result, much of the research in cogeneration systems is designed to figure out ways to use the excess heat energy beyond meeting space, water and process heating loads. Martin Engine Systems of Topeka, Kansas, a Caterpillar distributor, offers a 100-kw cogeneration system coupled to a Yazaki 15-ton absorption chiller. Re-Energy Systems is also coupling its 125-kw cogenerator with an absorption chiller. Thus the waste heat generated by a larger cogenerator during the summer can be used to power a large space-cooling system.

CHAPTER 7

Economics

Owning a power plant is philosophically appealing. Who wouldn't want to be independent of the local utility? Who wouldn't like to become more self-reliant, to become more insulated from future energy price hikes? But for most people the decision to invest in a small grid-connected power plant is based on sound economics, not philosophy.

A project's economic viability is judged by comparing its costs and benefits. Estimating project costs is much the simpler task. Equipment suppliers and contractors can provide specific cost estimates (although the risk of construction cost overruns remains). The terms of the project's financing are known before it is started. The term of the loan, the interest rate, the repayment schedule, the down payment and any other pertinent information will be specifically described in the contract before construction begins.

The project's benefits will be much harder to calculate. They consist primarily of guesses about the future. Even tax benefits may be uncertain because federal and state tax laws can change. The investor wants not only to maximize profit but also to minimize risk. A risk is a measure of uncertainty, and investments in small power plants present a greater array of uncertainties than other business investments. Will the technology function as expected? Will its availability factor meet manufacturers' claims? The cost of electricity from a machine operating at a 40 percent rather than an 80 percent availability factor is twice as high.

Will fuel be available in the quantity expected? Insolation varies relatively little from year to year but may vary significantly in any given month. This can affect project revenues if the qualifying facility (QF) operates under a time-of-day buyback rate. Wind regimes and stream flow rates can vary tremendously from year to year, greatly affecting the project's cash flow. Lean years must be expected along with the bounties of strong winds and heavy rainfall or snows.

What will the price of fossil fuel be in the future? The cogenerator using fossil fuels must not only project the future price the utility will pay but also the future price of fuel oil or natural gas consumed by the system. If the cogenerator is selling electricity to a utility whose marginal power plant is fueled by oil, while the cogenerator uses natural gas, the economics of the project will depend primarily on the relationship between future oil and gas prices. A cogenerator selling electricity to a utility with no future construction plans, that is, retiring its remaining oil- or gas-fired peaking plants to operate completely on coal or uranium, may find that profits in the early years will be eaten up by the higher cost of natural gas compared to coal in later years.

The greatest degree of uncertainty is in the level of future buyback rates. One study completed in mid-1982 for the city of Boulder evaluated the economics of various electric generation projects. It concluded that buyback rates under the Public Utility Regulatory Policies Act (PURPA) could range from slightly more than a penny to about 7.5 ¢, depending on the methodology used by the public service commission and the local utility. This range was possible even though all data on the cost of production to the utility was known. Imagine the possible variations when one is trying to project buyback rates 5 or 15 years into the future.

The investor often discovers the need to trade off profit for certainty. The uncertainty of future buyback rates can be diminished by negotiating a long-term contract with the utility. However, to gain such a contract, the QF must be willing to give up something. The price of gaining a long-term contract is often reduced profits. Long-term contracts might not have an escalator clause, but they might guarantee a base price for electricity. The base price in effect guarantees the investor a certain rate of return (assuming the technology operates as expected). In return the investor gives up the opportunity of making a much larger profit if future avoided energy costs should rise dramatically.

Insurance policies can also reduce some uncertainties. The manufacturer may be willing to guarantee the reliability of the technology, provided the buyer purchases a service contract. One cogenerator in New England took out an insurance policy to protect against the possibility that oil prices would rapidly *decline*. A wind-farm developer in Hawaii managed to purchase an insurance policy from Lloyds of London to insure that the wind would blow a certain portion of the time at a minimum speed.

Insurance policies are not cheap. An insurance company analyzes the risks involved and sets premiums to cover such probabilities plus to earn a healthy profit for itself. By spreading the risk over many ventures, the insurance company presumably can provide coverage at a lower price. But these premiums are expensive, which means that some projects may buy certainty at such a high price that the project is no longer economically viable.

A Simplified Guide to Energy Economics

The major determinants of the economics of independent production are: (1) the installed cost per kilowatt of capacity, (2) the capacity factor, (3) the financing terms, (4) the utility buyback rate or displaced cost of buying utility power and (5) the tax benefits.

The impact of each of these factors will be explored later in some detail. Here some simple rules of thumb are presented that can provide the potential producer with a good initial perspective.

The first step is to estimate the cost per installed kilowatt of capacity. For wind-power systems, this will be between $1,500 and $3,500, largely depending on the size of the system. Larger systems have lower per-kilowatt costs. For photovoltaic systems, the current installed cost is about $10,000 per peak kilowatt. Unlike the other technologies, this price is dropping rapidly and may be as low as $5,000 by 1985. Photovoltaic system costs do not significantly decline for larger installations. Hydroelectric system costs are more variable, because they depend not only on the size of the facility, but on the need for a dam, the spillway length, cost of the powerhouse and land acquisition costs for roadways as well. Costs range from $2,000 to $6,000 per installed kilowatt. The installed cost of a cogeneration plant is $1,000 to $2,000 per kilowatt.

To estimate the amount of electricity generated from these different technologies, a capacity factor, that is, the percentage of the year the generator will be operating at full rated capacity, must first be developed. An earlier section examined a chart developed by C. G. Justus (see figure 6–8) that estimated capacity factors for wind turbines at different rated wind speeds and in different wind regimes. Capacity factors for wind turbines will be in the 15 to 35 percent range. An average would be about 20 percent. For photovoltaics, you can assume that the system will receive the equivalent of full sun for 4 to 6 hours of each day, averaged over the year. That means a capacity factor of 16 to 25 percent. Assuming the owner has sized the hydroelectric system appropriately (that is, one that is based on the best year-round production of electricity) the capacity factor should be 75 to 90 percent. A cogeneration system can operate at full capacity 90 to 95 percent of the time.

Using these capacity factors, the conclusion is that a 1-kilowatt (kw) wind turbine will generate approximately 8,760 × 0.20 = 1,752 kilowatt-hours (kwh) per year. A 1-kw photovoltaic array will generate about the same. A 1-kw hydroelectric system will generate 8,760 ×

0.85 = 7,446 kwh per year. A 1-kw cogenerator will generate 8,760 × 0.95 = 8,322 kwh per year.

To discover the annual cost of owning the technologies, assume the owner has obtained 100 percent financing based on a 15-year, 12 percent loan. Thus the cost per $1,000 is $143 per year (principal plus interest). If the wind machine costs $2,500 per installed kilowatt, the annual carrying costs are $356. Assuming the turbine generates 1,732 kwh per kilowatt of installed capacity, the cost per kilowatt hour is 21¢. For photovoltaics, assuming a $10,000 installed cost, the cost per kilowatt hour generated is 82¢. A hydroelectric facility costing $4,000 per kilowatt will generate electricity at a cost of 7¢ per kilowatt-hour. The calculations for cogeneration are more complicated because fuel must be purchased for its operation.

What these calculations show is that without taking into account tax benefits or future increases in energy prices, the owner of a hydro-electric facility who is receiving more than 7¢ per kilowatt-hour from the utility (or who can use the electricity internally to displace electric-ity that would have cost more than 7¢ to purchase from the utility) would be making money from the beginning. For the wind machine owner to break even, the utility must be paying 21¢ per kilowatt-hour and for the photovoltaic owner, 84¢ per kilowatt-hour.

These figures should worry only those owners who cannot take advantage of the tax benefits realized from owning their own power plant. For example, the residential user receives a 40 percent tax credit for owning a renewable-based technology. Assuming the owner does not exceed the tax credit ceiling, this lowers the actual cost to the wind machine owner to 12¢ per kilowatt-hour and to the photovoltaic system owner to 50¢ per kilowatt-hour. In addition, the owner can benefit from the tax deductions derived from the interest paid on the loan. In the preceding example, of the $143 paid in the first year for each $1,000 borrowed, $119 is for interest. The homeowner can de-duct this from his or her taxable income. To the homeowner in the 40 percent tax bracket, every $100 in deductions translates into a $40 reduction in taxes paid to the federal or state government. During the first year, the benefits derived from the tax deductions for interest on the photovoltaic system for the buyer in the 40 percent tax bracket reduce the cost per kilowatt-hour produced by 30¢.

Tax considerations play a dominant role in determining the eco-nomic benefits of independent power. It is unfortunate that something as politically motivated and artificial should become a prime motivat-ing factor. Remember that the same and even greater tax benefits are received by the utility companies and are a significant factor in deter-mining the price they charge their customers.

Business tax benefits tend to be more favorable than residential tax benefits, because businesses can depreciate the equipment. That means they can deduct a portion of the cost of the facility every year. As will be explored in the next section, some homeowners are setting themselves up as corporations to take advantage of these benefits. It appears that this can be done only if all the electricity is sold to the utility rather than used internally.

Setting oneself up as a business is in keeping with Alvin Toffler's description of people becoming prosumers. In his best-selling book *The Third Wave,* Toffler describes homeowners who become producers as well as consumers. The person who buys a wind machine or a photovoltaic system should analyze such purchases differently than if he or she were buying a stereo or an automobile or refrigerator. The independent power system is an investment. The person who buys one should analyze its worth not only in psychological terms but in terms appropriate to any investment. Buying a wind machine is more akin to investing in a money market fund than to purchasing a refrigerator.

Opportunity Costs

Capital is a finite resource. The decision to invest in a power plant is really two decisions, one positive and the other negative. To invest in one project is not to invest in another. The benefits of one project must be compared to the lost opportunities of another investment. For example, $10,000 invested in a hydro project may generate a return of 10 percent on the investment after the loan and all other expenses are repaid. A similar investment in a corporate bond or money market fund may generate a 12 percent net return.

Investment in the money market fund (if the fund invests only in federally guaranteed securities) carries essentially no risk. Moreover, it is what economists call a *liquid investment.* It can be converted into cash at any time; checks may often be written on it even as the money is earning a risk-free return.

A small power plant carries a higher risk. Moreover, capital tied up in it cannot easily be converted to cash. Therefore, an investor may demand a higher return for this project than for a risk-free investment. When interest rates in federal securities were more than 15 percent in 1981, many investors were demanding a 25 percent return for investments in small power plants. In other words, they wanted their original investment repaid in less than four years.

The return the investor desires is called the *discount rate.* The discount rate is based on the inflation rate and the opportunity cost of money. Sometimes the term *real discount rate* is used. That is the discount rate after inflation has been taken into account. A real discount rate of 5 percent means the investor wants a return that is 5 percent greater than the inflation rate over the term of the investment.

To compare investment opportunities, one must take into account a wide variety of factors. For example, a money market fund invests in short-term securities. Thus the interest rate given to its investors fluctuates over brief periods. One may invest in the fund when interest rates are 15 percent but within a year they could drop to 10 percent or lower, as they did in late 1982. A small power plant investment is a long-term one. If one expects energy prices to rise faster than the general rate of inflation and if the buyback rate is based largely on the price of energy, then one might be justified in expecting to receive higher and higher returns in the future. Therefore, the slightly lower return at present might be justified on the basis of possibly very high returns in the future.

In making comparisons, one must be careful not to compare apples and oranges. The interest earned on a certificate of deposit at a bank or interest from a money market fund is taxable. If one is in the 40 percent marginal federal income tax bracket, the 12 percent return shrinks to 7.2 percent. Revenue gained from selling electricity to the utility is also taxable. But businesses are usually taxed at a lower rate than individuals. Moreover, the independent producer may be using some of the electricity on-site. The electricity not purchased from the utility is nontaxable income. If the utility charges 7¢ per kwh and the hydro plant displaces 1,000 kwh a month, the $70 normally paid to the utility is nontaxable income. (Put another way, if you are in the 30 percent tax bracket, the utility must pay you 30 percent more for your electricity than it charges you for its own before it makes it worth your while to sell it rather than consume it on-site.)

Investment decisions should also account for the *time value* of money. A dollar received a year from now is worth less than a dollar received today. At an annual inflation rate of 10 percent, a dollar received today and spent a year from now will have the purchasing power of only 91¢. Remember the misleading barrage of bank advertisements after the Individual Retirement Accounts (IRA) were established in 1981? They promised to make customers "millionaires" if they invested the maximum $2,000 each year for 30 years. The ads painted a picture that was more rosy than accurate, because they ignored the time value of money, that is, the reduction in purchasing power that inflation wreaks. True, anyone following the bank's advice would have a million dollars by retirement age. But a million dollars in the year 2015 would buy considerably less than it would have in 1981.

The loaf of bread in 2015 might cost $2.50. The average annual wage might be $50,000.

A more accurate analysis of the value of an IRA account would have taken inflation into consideration. Since long-term interest rates tend to be 2 to 3 percent higher than the long-term inflation rate, the actual rise in purchasing power for the investor would be 2 to 3 percent compounded over 30 years—significant but not sufficient to make the investor a 1981 millionaire.

The time value of money is taken into account by discounting each year's costs and benefits back to the present. Assume, for example, that the power-plant owner pays out an equal yearly loan payment of $3,000 for each of the first 10 years. What happens to the present value of that $3,000 for different inflation rates and for different time periods? Compare two cases (see table 7–1), one with a 15 percent inflation rate and the other with a 10 percent inflation rate. The table gives the present value of the $3,000 at the end of each year.

This comparison illustrates several points. Using a 15 percent inflation rate, $3,000 spent in the fifth year is worth only about $1,500 in present dollars. By the tenth year, the present value of the $3,000 has fallen to $742. In other words, an economic evaluation of this project takes into account that a revenue of $675 spent in the first year is equivalent to $3,000 spent in the tenth year. Note the dramatic difference in the present value of $3,000 based on the inflation rate

TABLE 7–1
The Time Value of $3,000

YEAR	ANNUAL INFLATION RATE OF 15%	ANNUAL INFLATION RATE OF 10%
1	$2,609	$2,727
2	2,268	2,479
3	1,973	2,254
4	1,715	2,049
5	1,492	1,863
6	1,297	1,693
7	1,128	1,539
8	981	1,400
9	853	1,272
10	742	1,157
Total	$15,058	$18,433

chosen, especially in the later years. Using a 10 percent inflation rate, the total present value of a stream of annual payments of $3,000 is $18,433. Discounting by 15 percent decreases the present value by 20 percent, to $15,058.

To compare costs and benefits properly, the revenue side of the equation should also be discounted. Thus, rising buyback rates should be discounted by some factor to be equated to the expenditures. With the assistance of a $15 pocket calculator and financial tables available in many economics and business textbooks, discounted cash flows can be developed.

The process of discounting for the time value of money is used not only to analyze the costs and benefits of small power plants but also to establish buyback rates. In many public service commission proceedings, the central dispute concerns the choice of the discount rate and its application to future plant construction costs or energy prices. The capacity credit portion of the buyback rate is the present value of avoided investments in future power plants. The capacity credit will be low if the cost of future power plants is low or if the discount rate is high. Since utilities want to minimize the capacity credit, they push back investment schedules, minimize the cost of future power plants and use a very high discount rate. As the above example illustrates, using a 15 percent rather than a 10 percent discount rate may lower the buyback rate offered QFs by 20 percent.

Utilities have other techniques to minimize PURPA rates. For example, construction cost increases for power plants have for the past decade been outrunning inflation. Therefore, the further into the future the construction takes place, the greater its present value. But many utilities ignore the historical data. They presume future construction cost increases will be below the inflation rate. This reduces the present value and reduces the buyback rates. Many arguments before public service commissions are about the choice of the discount rate used to deflate future dollars. Should it be the present rate of return the utility is permitted? The present rate of return is based on the interest paid on corporate bonds plus the return necessary to attract equity investors. This in turn is based on the inflation rate and long-term interest rates. Should the discount rate be based on the long-term interest rate or the short-term rate as illustrated in the rate of return? The difference in 1982 might be between 12 and 18 percent, making a dramatic difference in the present worth of investments ten years into the future.

When the time value of money is taken into account, benefits received in early years become disproportionately important compared to those received in later years. For example, $1,000 of tax credits that can be taken in the first year of operation may outweigh

$2,000 in revenue generated from rising buyback rates in the tenth year of operation.

Case Study: Wind Power

The economics of a specific wind-machine investment will be analyzed in some detail. Afterward, examples of hydropower, photovoltaic and cogeneration facilities will be examined in a more cursory fashion. The wind machine is a 10-kw machine rated to deliver 10 kw at 24-miles-per-hour (MPH) wind speed.

Costs can be divided between fixed costs and variable costs. Variable costs include insurance, taxes, metering fees and operation and maintenance (O&M). Assuming a simultaneous purchase and sales arrangement, metering costs are estimated to run $300 per year. O&M expenses are estimated to be 1 percent of the project costs or about

TABLE 7–2
Capital Costs
of a 10-kw-Rated Wind Generator

Wind turbine	$8,172
65-foot tower	3,480
Excavation and foundation labor	1,668
Field materials (concrete, electrical and so forth)	1,968
Field installation labor	372
Equipment rental	216
Transportation	420
Taxes (6% sales tax on materials)	817
Fees, permits and inspections	120
Contractor overhead and profit	2,352
Interconnection costs	3,500
Total	$23,085

SOURCE: These estimates are based on those developed by Arthur D. Little in 1980. See William C. Osborn and William T. Downey, *Near-Term High Potential Counties for SWECS*, p. 86. Insurance and interconnection costs were added. Costs were inflated by 20 percent to account for inflation between 1980 and 1982.

$230 per year. An exemption from property taxes is assumed. Insurance premiums cost $200 a year. Thus the recurring costs amount to $730 a year.

Project costs should also include the costs of negotiating the PURPA contract, the cost of obtaining Federal Energy Regulatory Commission (FERC) certification as a QF, the cost of monitoring the wind and any other incidental expenses. These costs often add up to several thousand dollars and can occur many months, or even years, before the project generates any revenue.

If this machine is rated at a 24-MPH wind speed and operates in a 12-MPH average wind-speed regime, it will have a capacity factor of 20 percent. Thus it would have an annual output of 17,520 kwh per year (8,760 × 10 kw × 0.20).

An accurate evaluation takes into account tax credits, the interest rate on the loan to finance the system, the timing of future energy price increases and the timing of the output of the wind machine. Then a cash flow is developed for each year of the project. From this data one can estimate the return on investment.

Assume, for example, that the system cost of $23,085 is financed by a ten-year, 15 percent loan. Assume further that 80 percent financing is available for the long-term loan. The owner puts up 20 percent of the installation cost in cash.

The first year, the project generates revenue based on the estimated 17,520 kwh annual output, depending on how long it takes to install the wind machine. A more accurate cash flow examination would take into account the two- to four-month delay between the time the financing is taken out and the time the machine actually begins to generate power. The down payment for the long-term loan comes to about $4,600. Annual payments on the remaining $18,500 loan come to about $3,600. The first year interest is $2,700, which is of course deductible, in addition to maintenance expenses of $730 per year, also deductible. (The interest deduction will be less for each succeeding year.)

The second year of the project has the same expenditures as the first except for the down payment. This is also true for the remaining years of the project.

Tax Benefits

One further step remains to evaluate the economics of this project. Tax incentives must be included. Tax incentives for small-scale power production are available for residential and commercial enterprises. Different levels of benefits are available for different technologies and for business and residential systems.

Tax deductions are distinct from tax credits. The former is taken against gross income, while the latter is taken against tax liability. This makes the latter much more attractive. For example, suppose the individual homeowner is in the 35 percent tax bracket (combined federal and state taxes). A $1,000 tax deduction reduces gross taxable income by $1,000. This reduces tax liability by $350. The benefit of tax deductions is directly related to the tax level of the individual. If this homeowner were in the 50 percent tax bracket, the deduction would be worth $500. If his or her tax bracket were 15 percent, the deduction would be worth only $150. Tax deductions are useful only if one itemizes deductions on the income-tax form.

Tax credits are taken directly against tax liability. A $1,000 tax credit reduces taxes by $1,000. They can be taken even if one does not itemize deductions.

Both deductions and credits can usually be carried forward (and sometimes can be carried backward). Thus if one pays no taxes this year, but paid taxes last year or will pay them next year, the tax benefits can be applied in those years.

Small power plants are eligible for four types of tax benefits: investment tax credit, energy tax credit, depreciation and interest deductions.

The investment tax credit (available only to businesses) of 10 percent can be used by all businesses to offset investments in machinery. It is not restricted to investments in energy-related technologies.

Energy tax credits are available for all electric generation technologies covered in this book. Cogenerators are eligible, however, only if they are predominantly fueled by a renewable resource such as methane, solid waste or solar energy. Energy tax credits are available for residences as well as businesses. The level of the tax credit for businesses varies by technologies. Photovoltaic or wind technologies qualify for a 15 percent tax credit. Hydroelectric plants receive an 11 percent credit. Until 1982 there was a 10 percent tax credit for cogeneration plants, but this is available now only for those fueled by renewable resources such as methane or solid waste. There is no credit for cogeneration plants for residences. The residential tax credit for solar and wind is 40 percent at the federal level on a maximum investment of $10,000 ($4,000 maximum credit). Business tax credits have no ceilings.

One can deduct the depreciation of the equipment used in a small power system. This deduction is allowed under the presumption that the business is setting aside money to finance a new machine. The 1982 tax law renamed depreciation *accelerated cost recovery system* (ACRS). Depreciation is unavailable for residential ownership. However, it is available for technologies located on residential property but owned and operated by a commercial enterprise, which a homeowner could conceivably become. The 1982 tax law requires the power plant owner to reduce the basis for depreciation by 50 percent of the energy and investment tax credits. Thus if the equipment costs $100,000 and qualifies for 25 percent combined tax credits, the depreciation can be taken against a total equipment cost of $87,500.

The interest paid on a loan can be deducted. Given the time value of money, this provides an incentive to pay the interest up front. If one is in the 30-percent tax bracket, then 30¢ on the dollar paid to interest is actually being paid by Uncle Sam or the state or local government.

Many states provide additional tax benefits. Sometimes these can be used in addition to the regular federal tax credits, sometimes they are not additive. Kansas, for example, has a 30 percent tax credit for wind-electric machines that is taken in addition to the 40 percent federal income tax credit. California provides a state tax credit for the cost of any wind electric system that exceeds $12,000 and is not installed on a residential premise. This credit has no limit and is not reduced by any federal tax credits taken. Those who install a wind electric system that costs less than $12,000 or that is installed on residential premises are eligible for a 55 percent tax credit for the system's cost. This credit has a limit of $3,000 and is reduced by the federal residential tax credit. California taxpayers can also depreciate the entire cost of a wind electric system in one, three or five years instead of taking the state tax credit, or they can take the state credit and depreciate all costs in excess of the credit in three years.

Continuing with the wind machine example, the residential owner is eligible for the 40 percent federal tax credit on a maximum investment of $10,000. This provides a direct credit against taxes of $4,000. If the resident is in the 30 percent tax bracket, $3,430 in deductible expenses (first year interest of $2,700 and operating expenses of $730) equals $1,029 in reduced taxes, bringing the total first year tax reduction to $5,029. The energy tax credit is a one-time benefit. The interest deductions continue for the life of the loan.

The first year there is an outflow of about $9,100, generating a tax benefit of about $550 plus the tax credit of 40 percent of the first $10,000, or $4,000 for a total of $4,550. The second year the expenditure is $4,500 in loan payments. The tax benefit is $550 for a net expenditure of $3,900. That remains true for the remainder of the loan. If we rearrange these costs into net expenditures, the ten years give the following: the first year is $4,600, while subsequent years are $3,900 each for a total of $39,700 for ten years. Over the ten years the system will generate 175,200 kwh for a kwh cost of 22.7¢ ($39,700 ÷ 175,200 kwh).

Sensitivity Analysis

Any economic analysis should change key variables to examine how the results could vary. Then the most likely combination of variables should be chosen for the final analysis, with a likely range of values included in the final analysis as well. This is called a *sensitivity analysis*.

The situation painted above may actually be considered a worst-case scenario. It assumes a modest wind regime, a very short system life, a residential application with a ceiling on the tax credits, relatively high interest rates and no equipment salvage value. A worst-case scenario is an important antidote to Pollyannaish optimism. But a more realistic assessment of the project's economics would alter key variables.

One of the key variables here is wind speed. If the site had a 14-MPH rather than a 12-MPH wind speed, the annual electrical output would increase to 24,528 kwh. The kilowatt-hour cost over the ten-year period drops to 14¢.

This evaluation used an extremely short life expectancy for the wind machine. A more likely time period would be 20 years. Assuming a 12-MPH mean wind speed, the system will generate 350,400 kwh over 20 years. The additional 10 years will mean additional recurring costs (for maintenance) of $7,300 (10 × $730), raising total expenditures to roughly $50,000. The cost per kilowatt-hour is then about 14¢.

The system will also be eligible for state tax credits, depending on its location. New York and California have a 55 percent tax credit for wind machines. This credit will lower the system cost by about $2,300, reducing the cost of electricity by 2¢ per kilowatt-hour (in a 12-MPH wind regime).

Will the machine have any resale or salvage value? If a resale value of $2,000 can be expected in 20 years, the cost of electricity is lowered again by about 2¢ per kilowatt-hour. A similar reduction would occur if the O&M costs dropped by $100 a year over the life of the system, or if the metering costs dropped by a like amount.

If the wind machine has a built-in synchronous inverter and is located very near the utility line, the interconnection costs could drop by $2,500 to $3,000. This reduces the cost per kilowatt-hour by an additional 2¢ to 3¢.

Assuming all the most favorable factors, a wind machine operating for 20 years in a 12-MPH wind regime in California or New York would generate electricity at a cost of about 6¢ per kilowatt-hour. Assuming that interest rates for loans will also drop so that the original loan can be refinanced at a lower rate, the figure will drop still further.

This ends the examination of the expenditure part of the equation.

The key to the profitability of the project is the buyback rate. Today, buyback rates vary from less than 2¢ to about 9¢ per kilowatt-hour. In this example, assuming a 20-year system life, the total revenue will range from $7,000 to $31,500. Only under the most optimistic expenditure assumptions would the high end of this range of buyback rates generate a net profit for the project. However, the wind machine owner expects electric prices to rise, a rise that should be accompanied by an increase in the buyback rate. The profit of many of these projects is based on the expectation of future rises in avoided costs to the utility. For example, if the initial buyback rate is 5¢ per kilowatt-hour but escalates at an annual rate of 10 percent, the total revenue received over the 20-year period jumps from $17,500 to $50,000. By the twentieth year, the wind machine owner will be receiving 34¢ per kilowatt-hour. Figure 7–1 shows the relationship of energy inflation rates to the constant cost of the output of the wind machine.

The wind machine owner may be using all or most of the electricity on-site, or may have a net billing arrangement with the utility that reverses the meter. In that case, there may be no metering charges and few or no interconnection charges, and the price the producer receives for the electricity is the retail price. In most parts of the country, despite

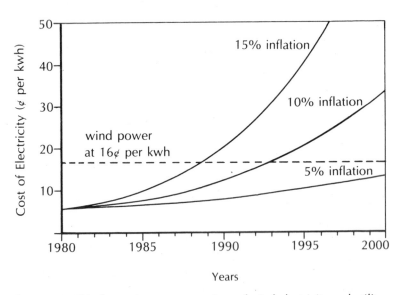

Figure 7–1: This figure shows a comparison of wind electricity and utility costs. The cost of the wind power assumed a $5,000 fee for purchase and installation, in addition to insurance and maintenance costs of $200 a year during a lifetime of 20 years. *Adapted by permission of Cheshire Books from* The Wind Power Book © *1981 by Jack Park.*

PURPA's emphasis on marginal costs, utilities are paying less than their retail rate to QFs. Thus the wind machine owner will be generating a greater profit by consuming on-site. If this is the case, it also pays the wind machine owner to size the machine to use all of its consumption on-site. If the buyback rate is higher than the retail rate by more than the marginal tax bracket of the producer, then it encourages the sizing of the wind machine to extract a maximum amount of energy from that site.

The economics of the project change if they are examined as a commercial enterprise. Commercial projects qualify for investment and energy tax credits and accelerated cost recovery tax deductions besides the deductions for interest. Corporations often pay taxes in the 40 percent bracket. The machine may be owned by a group of investors called a *limited partnership*. In that case, the investors can take the tax benefits directly on their individual income tax returns. Limited partnerships are called tax shelters because they are a way to avoid paying taxes on income generated from some other enterprise or personal wages. The investors in these shelters are usually in the 50 percent tax bracket (higher if state taxes are taken into account). Thus deductions are worth much more to them than to the homeowner in the 30 percent tax bracket. From the previous wind system example assume a combined energy and investment tax credit of 25 percent (on the noninterconnection costs), a 40 percent business marginal state and federal income tax and a five-year ACRS. The first-year cash outflow of $8,930 (down payment of $4,600, mortgage payments of $3,600, operating expenses of $730) is nearly offset by the applicable tax benefits. Deductions of $6,480 (interest of $2,720, depreciation of $3,030, operating expenses of $730) generate $2,592 of tax benefits at the 40 percent marginal tax rate. This amount coupled with the allowable tax credit of $5,771 ($23,085 × 0.25) results in a total first-year tax benefit of $8,362, which almost entirely offsets the first-year cash outlay. Of course, the tax benefits decline in each ensuing year as a greater portion of each year's loan payments goes toward the principal.

The total expenditures over the 10-year loan repayment are $26,650. Assuming a 10-year life for the system, the cost per kilowatt-hour would be 15¢. Assuming a 20-year life, the cost would rise to about $33,000, but the kilowatt-hour cost drops to about 9¢. State tax credits, lower O&M costs and better loan terms or reduced interconnection costs would further lower this cost. Larger wind machines in windier regimes will be able to generate electricity at a much lower cost.

Ned Coffin, president of Enertech Corporation, a manufacturer of small wind systems based in Norwich, Vermont, describes in an article in *Alternative Sources of Energy* magazine the following situation where a Vermont resident purchased a 4-kw wind machine.[1] Instead of wiring the system to his house, he had the company wire it so that 100 percent of the output went to the utility.

The homeowner was going to create a subchapter S corporation. This type of corporation can be incorporated with little paperwork and allows the owner or owners to consolidate the income and losses of a business with their personal income, something that stockholders of regular corporations cannot do. Yet it still retains most of the other advantages of a true corporation, including a limited liability for the owners. In a subchapter S corporation, the owner can take a tax benefit granted to the business directly on his or her personal income tax.

The system cost $16,000. The enterprising Vermonter qualified for an immediate 10 percent federal investment tax credit and a 15 percent wind energy tax credit. In addition, the business qualified for an immediate Vermont business tax credit of $3,000. Moreover, the business can take off the first year's depreciation of 15 percent (even if the machine is purchased on December 31). On top of these tax breaks, the homeowner-owned corporation gets the income from the utility purchases of electricity (less the taxes paid on this earned income).

Tables 7–3 and 7–4 indicate the cash flow and the payback periods for the investment. Notice that the owner in the 50 percent tax bracket repays the initial investment in 5 years, whereas someone in a more typical 30 percent bracket would have a 6.5-year payback.

Coffin notes that his customer planned to abandon the corporation and to rewire the system to meet his household requirements once the system had paid for itself and all depreciation allowances had been used up. He estimated that at that time he would be getting a much higher return by using electricity internally than by selling it to the utility company and paying income tax on the proceeds.

Assuming a 10 percent annual increase in buyback rates and an owner in the 50 percent tax bracket, the initial investment of $16,000 will be recouped in about five years. In other words, the owner is getting about 20 percent return after taxes. If the owner is in a lower tax bracket, the payback period is slightly longer.

This example assumes an initial investment of $16,000; however, that investment need not mean a $16,000 up-front cash outlay. For example, the owner might pay $4,000 down and borrow the remaining $12,000. The tax credits can still be taken against the full $16,000 and the depreciation can be taken on $14,000. Thus, by putting a cash down payment of $4,000 plus paying the debt service on the remaining $12,000, the independent power producer can generate $8,410 in income and tax benefits the first year. Assuming all the tax benefits can

TABLE 7-3
Cash Recovery

		CUMULATIVE DOLLARS RETURNED
Initial Investment—$16,000		
Federal Energy and Investment Tax Credits (25%)	$4,000.00	
Immediate Vermont Business Tax Credit ($3,000 max)	3,000.00	
1st year depreciation (15% × $14,000 × 50% tax bracket)	1,050.00	
1st year income (9,200 kwh @ $0.078 less 50% taxes)	360.00	
Total 1st year cash recovery	$8,410.00	$ 8,410.00 (52.6%)
2d year depreciation (22% × $14,000 × 50% tax bracket)	$1,540.00	
2d year income (9,200 kwh @ $0.086 less 50% taxes)	395.00	
Total 2d year cash recovery	$1,935.00	$10,345.00 (64.7%)
3d year depreciation (21% × $14,000 × 50% tax bracket)	$1,470.00	
3d year income (9,200 kwh @ $0.094 less 50% taxes)	430.00	
Total 3d year cash recovery	$1,900.00	$12,245.00 (76.5%)
4th year depreciation (21% × $14,000 × 50% tax bracket)	$1,470.00	
4th year income (9,200 kwh @ $0.104 less 50% taxes)	480.00	
Total 4th year cash recovery	$1,950.00	$14,194.00 (88.7%)
5th year depreciation (21% × $14,000 × 50% tax bracket)	$1,470.00	
5th year income (9,200 kwh @ $0.114 less 50% taxes)	525.00	
Total 5th year cash recovery	$1,995.00	$16,190.00 (101.2%)

NOTE: Reprinted from *Alternative Sources of Energy* magazine, issue 59. For subscription information, write to them at 107 S. Central Ave., Milaca MN 56353.

TABLE 7–4

Cash Recovery Calculations
(Wired to Utility, then House, vs Wired to House)

I. ASSUMPTIONS	WIRED TO UTILITY	WIRED TO HOUSE
Installed Cost	$16,000	$16,000
Federal Tax Credits	$ 4,000	$ 4,000
State Tax Credits	$ 3,000	$ 1,000
Net Cost after Tax Credits	$ 9,000	$11,000
Depreciation*	$14,000	- 0 -
Output (kwh)	9,200	9,200
Price/kw	$ .078	$ .078
1st Year's Earnings/Savings	$ 718	$ 718
Annual Rate Increase	10%	10%

	OWNER'S TAX BRACKET		
II. WIRED TO UTILITY, THEN HOUSE	50%	40%	30%
Net Cost after Tax Credits	$9,000	$9,000	$9,000
Taxes Saved by Depreciation	$7,000	$5,600	$4,200
Net Cost to Owner	$2,000	$3,400	$4,800
1st Year's Earnings, Less Taxes	(359)	(431)	(503)
5 Years' Earnings, Less Taxes–Assuming 10% Annual Rate Increase	$2,190	$2,630	$3,068
Balance Remaining, 5th Year	-0-	$ 770	$1,732
6th Year's Cash Savings–Assuming 10% Annual Rate Increase†	$1,149	$1,149	$1,149
Years to Cash Recovery	5	5.5	6.5
III. WIRED TO HOUSE			
Net Cost after Tax Credits	$11,000	$11,000	$11,000
1st Year's Cash Savings†	(718)	(718)	(718)
5 Years' Earnings–Assuming 10% Annual Rate Increase†	$ 4,380	$ 4,380	$ 4,380
Balance Remaining, 5th Year	$ 6,620	$ 6,620	$ 6,620
6th Year's Cash Savings–Assuming 10% Annual Rate Increase†	$ 1,149	$ 1,149	$ 1,149
Years to Cash Recovery	10	10	10

NOTE: Reprinted from *Alternative Sources of Energy* magazine, issue 59. For subscription information, write to them at 107 S. Central Ave., Milaca MN 56353.

* Can be used to offset taxes on corresponding amount of income from other sources.

† Cash savings only are shown. For persons to whom $1,000 saved is equal to $1,500–$2,000 earned, the value of these savings is obviously higher.

be used in that year (or carried forward to be used in later years), the return on cash outlay is more than 100 percent the very first year.

In my opinion, the next steps in the economic evaluation for both business and residence are to develop a 20-year cash flow and to discount the various benefit and expenditure streams to their present value. This weights the evaluation to favor those benefits received early (e.g., tax credits) versus those received later (e.g., rising buyback rates). Earlier it was noted that at a 10 percent escalation rate, a 5¢-per-kilowatt-hour buyback rate in 1983 becomes a 34¢-per-kilowatt rate in 2003. But we must also discount that back to the present, reducing its value to current investors considerably. This is why investors will often sign contracts with utilities that give them higher than normal buyback rates immediately, possibly 125 to 150 percent of the regular rate, and then reduce that rate over the next few years so that in later years they are receiving less than the regular rate. The discounted benefits are then compared to the discounted costs and an internal rate of return is computed. If the return is equal to or greater than that desired by the investor, the project is attractive.

As the reader can see, an examination of any project is complete with arithmetic exercises, hard work and educated guesses about the future. The reward is that the prospective power plant owner will have a much clearer idea of the range of possible outcomes of the project and its value.

Hydropower

Construction-cost estimates for hydropower systems may be subject to more variation than estimates for wind-power systems. Part of the reason is that regulatory delays can increase construction costs through inflation. Another is that regulations can impose conditions on flow rates and require fish ladders or certain engineering designs that increase the project's costs. The costs for hydroelectric facilities will vary significantly depending on whether there is an existing dam, whether used equipment is installed, how far the power plant is from the utility and the cost of interconnection equipment. For this example, assume a sufficient flow rate and head to support a 50-kw plant. The installation cost, including switch gear and interconnection equipment, is assumed to be $3,000 per kilowatt for a total cost of $150,000. The facility is assumed to operate at a 60 percent capacity factor, typical of facilities lacking storage ponds. At a 60 percent capacity factor, this plant will generate 262,800 kwh annually.

Hydro facilities last more than 40 years. Here assume a 25-year life in which the system will generate 6.5 million kwh. Assuming its $150,000 installation cost is financed at 15 percent over 10 years, the total principal and interest paid out is slightly more than $290,000. Assuming operation and maintenance expenses of 1 percent of the system's cost and insurance of $1,000 per year, the total expenditures over 25 years come to about $350,000. The cost per kilowatt-hour generated is 5.4¢. Interest deductions and tax credits will lower this figure to 2¢. The reader can understand why prospectors searched the nation in the early 1980s looking for viable hydro sites with existing dams.

Cogenerators

An economic analysis of cogeneration plants is more complicated because one must take into account rising fuel prices as well as the other cost components. For this analysis a Re-Energy system is used. The system uses a 140-horsepower (HP) Mack Truck engine operated at 70 HP. The engine drives a 50-kw generator. The system is designed to operate 100 percent of the time to meet the base load. Under full load the system converts about 38 percent of its energy burned into electricity. An additional 58 percent of the primary energy is recovered as useful thermal energy.

The installed cost of the system is $80,000. Assuming a 12 percent, five-year loan on the system, the yearly payments are $21,143.

For every one million Btu of primary energy burned in the Re-Energy system, 380,000 Btu of electric power are generated, or 111 kwh and 580,000 Btu of useful thermal energy are recovered. Using a 1983 average national cost of natural gas of $7 per million Btu (70¢ per therm), the cost of generating a kilowatt-hour of electricity is 6.31¢ if we assumed no heat were recovered. Given that the efficiency of the heat recovery compares favorably with the efficiency of a commercial boiler (although it is about 10 to 15 percent lower than the efficiency of a well-tuned residential furnace), one could be justified in viewing the electrical output as having no additional cost. The inherent economics of a cogeneration system is that it provides two products for the price of one. And one of those products—electricity—has a very high value to the user. Seven cents per kilowatt-hour, the typical cost of electricity, translates into more than $20 per million Btu, nearly three times the cost of natural gas or fuel oil.

If this system were run 100 percent of the time, it would produce 50 kw × 8,760 hours = 438,000 kwh per year plus 2.3 billion Btu of thermal energy.

We can now estimate the payback. If all the electricity were used internally and displaced 7¢ per kilowatt-hour, the return would be $30,660. Thus the net operating profit on the system is $9,000. (The company provides a warranty for the system for five years and takes care of all maintenance, including oil changes and valve adjustments.)

With tax benefits added in, the return is even more attractive. The residential owner receives no tax benefits. Given the scale of even the smallest cost-effective cogeneration systems and the fact that tax benefits are available only to commercial enterprises, these systems will be predominantly installed in commercial structures or under a corporate umbrella for an association of homeowners or apartment house dwellers.

The 10 percent investment tax credit and 10 percent credit for cogeneration (which expires in 1983 but may be extended by Congress) gives a first-year tax credit of $16,000. The depreciation in the first year is 15 percent. Assuming a 40 percent tax bracket for the business, this is equivalent to a 6 percent tax credit. Thus the business gets more than $20,000 in tax benefits the first year as well as the $9,000 net income from the electricity.

In this case assume equal efficiencies for the cogeneration system and the commercial boiler. Therefore, the natural gas consumed would have been consumed anyway for space heating and domestic hot water.

There are many variables to watch for when installing a cogeneration system. If the boiler were 80 percent efficient compared to the thermal energy recovery efficiency of 58 percent for the unit in the example, one would have to calculate in the additional natural gas burned to generate the electricity. If the system operated only 50 percent of the time, the operating cost for generating electricity would be 6.3¢ per kilowatt-hour. Thus the 7¢ displaced generates less than a penny return per kilowatt-hour.

This example assumed all electricity was used internally. This is one reason the system was sized to generate 50 kw. If the buyback rate of the utility were 7¢ or over, the owner would probably install a larger system (depending on the use to which the waste heat could be put) and export even more to the utility. But in most parts of the country, the buyback rate will only be a fraction of the rates the utility charges the customer. If the buyback rate were 5¢ per kilowatt-hour and the system operated 100 percent of the time, the operating profit would drop to about $6,500.

The comparative cost of the fuel used for the cogeneration system and for the central power plant is an extremely important variable. This example assumed a cost of $7 per million Btu for natural gas. If the utility is meeting increased demand and has oil-fired capacity, the buyback rate should be 5¢ to 7¢ per kilowatt-hour. The lower the cost of natural gas, the more attractive the cogeneration installation will be. If, on the other hand, gas prices are high and the utility is coal fired with no plans for any future capacity additions, the cogeneration system could not be economically justified on the basis of generating electricity for sale.

Cogeneration plants have an advantage that wind power and photovoltaics and, to a certain extent, hydropower do not have. They can and should qualify for capacity credits. They can meet any reliability standard utilities offer, and they can be operated so that maintenance takes place in the off-peak periods, allowing them to qualify for time-of-day energy credits. Thus cogeneration facilities should be able to qualify for the highest buyback rates offered. Capacity credits can add 1 to 2¢ per kilowatt-hour to the buyback rate.

However, capacity credits or firm energy credits usually require long-term contracts. If there is no escalator clause to the contract, the cogeneration plant fired by natural gas will be in a particularly difficult situation as gas prices continue to rise dramatically. If the local utility has peaking plants fired by oil, the rising gas prices will possibly increase the generation cost of electricity on-site more rapidly than avoided energy costs increase at the utility level. If the utility is switching to coal and nuclear plants and is retiring its oil- or gas-fired peaking plants, the situation could become worse for the independent power producer. If the natural gas price increases at a faster pace than do electric rates, the project will experience a loss in the later years.

Cogeneration systems can also be used as load-leveling devices. They can be turned on during peak periods of the day or year to reduce demand charges. Only commercial customers now pay demand charges. These can represent 50 percent of the total utility bill. In the case of peak-shaving or load-leveling uses, the cogeneration system need not be designed for long periods of use between maintenance. It can have a higher number of revolutions per minute (RPM) and will cost considerably less than the model used in this example.

Photovoltaics

A small household photovoltaic (PV) system can use the electricity generated to meet the household load, or it can sell it to the utility or store it in batteries or in hot water tanks. Whichever strategy is chosen will in large part be based on the buyback rate, the cost of storage and other factors.

The size of a residential system in the next few years will probably be influenced by the ceiling on the residential solar tax credit. The 40 percent federal tax credit can be taken only on the first $10,000. That price is currently sufficient to install a 1-peak-kilowatt system. In a typical location, that system will generate about 1,700 kwh annually. Since the typical American household consumes 7,500 kwh a year, one might be tempted to assume that the entire output could easily be used on-site. But remember, the electricity is not generated from the PV device at precisely the time it is needed by the internal load. Therefore, during parts of the year, the household will have to store, dump or export the surplus electricity. If the surplus coincides with the utility's air-conditioning peak, the PURPA buyback rates may be attractive. For a household system, the metering costs, back-up rates and interconnection charges become primary factors in encouraging or discouraging utility sales. Net billing arrangements where the household can use the original watt-hour meter are most attractive.

If the utility charges a high back-up rate (possible under an excess sales arrangement) or if its buyback rate is too low, the household owner may decide to store the electricity on-site. This can be done in batteries. But it can also be done by converting the electricity to hot water. Each 100 kwh of extra capacity per month is enough to raise the temperature of 20 gallons of water by about 70°F a day, enough to meet one person's hot water needs. The economic benefit derived from this depends on the cost of heating hot water ordinarily. The additional capital costs for the PV owner might consist of a transfer switch to direct the electricity to the storage tank when no load is present and possibly a larger water storage tank to prevent overheating of the water.

Assuming a system cost of $10,000, operation and maintenance costs of 1 percent or $100 per year and a 10-year, 15 percent loan, the total expenses over 20 years will be $21,000. Interest payments are about $9,400. The residential tax credit generates $4,000 in the first year of operation. (California and New York have 55 percent tax credits, lowering the price still further.) Assuming a 35 percent combined state and federal income tax bracket, the tax deductions reduce the expenditures by another $3,150, reducing total expenditures to $13,850. Since the system will generate 34,000 kwh over 20 years, the total cost per kilowatt-hour is slightly over 40¢.

This is too high for most people. But remember that the revenue for this household will come primarily from displacing utility-purchased electricity. The retail rate of electricity in some parts of the country is now 16¢ per kilowatt-hour and is increasing on average 10 to 15 percent a year. Meanwhile, the cost of PVs is declining. And there is the distinct possibility that cogeneration-PV systems will be developed. These will capture the thermal energy from the solar collectors as well as the electrical energy, increasing the revenue further. Imagine charting this on a graph. The two lines of reducing cost and increased retail price of electricity are expected to intersect in some parts of the nation by as early as 1985.

Obviously, a commercial PV system would be larger than one for a household. To evaluate such a system, assume a 20-kw peak array with an installed cost of $9,000 per peak kilowatt. The total loan is for $180,000. Assume a 10-year, 15 percent loan term. The cost for the first year will be $34,800, of which $26,400 is interest. Assuming operation and maintenance costs of 1 percent of system costs, the additional expenditures will be $1,800 a year. The total expenditures over the 20-year life of the system will come to $380,000. Total electricity generated over 20 years is 680,000 kwh. Total cost per kilowatt-hour generated in gross terms is thus 56¢.

The system is eligible for a 25 percent tax credit with no ceiling. It is also eligible for ACRS over five years. And the interest payments, as with the residential owner, are tax deductible. Assuming the facility is owned by a limited partnership with each of the partners in the 50 percent tax bracket, these tax benefits total $215,000. The cost per kilowatt-hour drops to 24¢. If the system is located in the higher insolation areas of the nation (e.g., the Southwest), the output from the 20-kw array could increase by 20 percent, reducing the cost per kilowatt-hour to less than 20¢.

In summary, an economic analysis is the single most important step *before* investing in small power production. This is the step in which all the data previously gathered is combined, where the cost data, the operational data, the PURPA rates, the internal-load-curve data and evidence of future buyback rates and financing terms are put together to provide a picture of the cash flow generated from the project. The economic analysis should project the cash flow, that is, the expenditures and revenues generated in each year, or each month, so that the owner will be able to identify those periods, especially during the early years, where more money will be going out than coming in. A sensitivity analysis can identify the factors that could substantially change the picture (e.g., changing reliability factors, future buyback rates and changing interest rates). The economic analysis should also provide several future scenarios, using a low, middle and high range for various factors and, if possible, applying a risk factor to each scenario.

Completing this task provides a good picture of what can go right, what can go wrong and what the venture looks like overall. Sometimes an economic analysis clearly cautions against investing. In most cases it defines the parameters or range of possible returns and the likelihood that the project will fall within that range. The decision to go or not is still an individual one. All the data and analysis in the world will not persuade an individual used to investing in federally guaranteed short-term securities suddenly to embark on a novel and risky venture. It does allow you to look before you leap.

Appendix 1

PURPA Sections 201 and 210

SEC. 201. DEFINITIONS.

Section 3 of the Federal Power Act is amended by inserting the following before the period at the end thereof:

"(17) (A) 'small power production facility' means a facility which —

"(i) produces electric energy solely by the use, as a primary energy source, of biomass, waste, renewable resources, or any combination thereof; and

"(ii) has a power production capacity which, together with any other facilities located at the same site (as determined by the Commission), is not greater than 80 megawatts;

"(B) 'primary energy source' means the fuel or fuels used for the generation of electric energy, except that such term does not include, as determined under rules prescribed by the Commission, in consultation with the Secretary of Energy —

"(i) the minimum amounts of fuel required for ignition, startup, testing, flame stabilization, and control uses, and

"(ii) the minimum amounts of fuel required to alleviate or prevent —

"(I) unanticipated equipment outages, and

"(II) emergencies, directly affecting the public health, safety, or welfare, which would result from electric power outages;

"(C) 'qualifying small power production facility' means a small power production facility —

"(i) which the Commission determines, by rule, meets such requirements (including requirements respecting fuel use, fuel efficiency, and reliability) as the Commission may, by rule, prescribe; and

"(ii) which is owned by a person not primarily engaged in the generation or sale of electric power (other than electric power solely from cogeneration facilities or small power production facilities);

"(D) 'qualifying small power producer' means the owner or operator of a qualifying small power production facility;

"(18) (A) 'cogeneration facility' means a facility which produces—

"(i) electric energy, and

"(ii) steam or forms of useful energy (such as heat) which are used for industrial, commercial, heating, or cooling purposes;

"(B) 'qualifying cogeneration facility' means a cogeneration facility which—

"(i) the Commission determines, by rule, meets such requirements (including requirements respecting minimum size, fuel use, and fuel efficiency) as the Commission may, by rule, prescribe; and

"(ii) is owned by a person not primarily engaged in the generation or sale of electric power (other than electric power solely from cogeneration facilities or small power production facilities);

"(C) 'qualifying cogenerator' means the owner or operator of a qualifying cogeneration facility;

"(19) 'Federal power marketing agency' means any agency or instrumentality of the United States (other than the Tennessee Valley Authority) which sells electric energy;

"(20) 'evidentiary hearing' and 'evidentiary proceeding' mean a proceeding conducted as provided in sections 554, 556, and 557 of title 5, United States Code;

"(21) 'State regulatory authority' has the same meaning as the term 'State commission', except that in the case of an electric utility with respect to which the Tennessee Valley Authority has ratemaking authority (as defined in section 3 of the Public Utility Regulatory Policies Act of 1978), such term means the Tennessee Valley Authority;

"(22) 'electric utility' means any person or State agency which sells electric energy; such term includes the Tennessee Valley Authority, but does not include any Federal power marketing agency."

SEC. 210. COGENERATION AND SMALL POWER PRODUCTION.

(a) COGENERATION AND SMALL POWER PRODUCTION RULES. —Not later than 1 year after the date of enactment of this Act, the

Commission shall prescribe, and from time to time thereafter revise, such rules as it determines necessary to encourage cogeneration and small power production which rules require electric utilities to offer to —

(1) sell electric energy to qualifying cogeneration facilities and qualifying small power production facilities and

(2) purchase electric energy from such facilities.

Such rules shall be prescribed, after consultation with representatives of Federal and State regulatory agencies having ratemaking authority for electric utilities, and after public notice and a reasonable opportunity for interested persons (including State and Federal agencies) to submit oral as well as written data, views, and arguments. Such rules shall include provisions respecting minimum reliability of qualifying cogeneration facilities and qualifying small power production facilities (including reliability of such facilities during emergencies) and rules respecting reliability of electric energy service to be available to such facilities from electric utilities during emergencies. Such rules may not authorize a qualifying congeneration facility or qualifying small power production facility to make any sale for purposes other than resale.

(b) RATES FOR PURCHASES BY ELECTRIC UTILITIES. — The rules prescribed under subsection (a) shall insure that, in requiring any electric utility to offer to purchase electric energy from any qualifying cogeneration facility or qualifying small power production facility, the rates for such purchase —

(1) shall be just and reasonable to the electric consumers of the electric utility and in the public interest, and

(2) shall not discriminate against qualifying cogenerators or qualifying small power producers.

No such rule prescribed under subsection (a) shall provide for a rate which exceeds the incremental cost to the electric utility of alternative electric energy.

(c) RATES FOR SALES BY UTILITIES. — The rules prescribed under subsection (a) shall insure that, in requiring any electric utility to offer to sell electric energy to any qualifying cogeneration facility or qualifying small power production facility, the rates for such sale —

(1) shall be just and reasonable and in the public interest, and

(2) shall not discriminate against the qualifying cogenerators or qualifying small power producers.

(d) DEFINITION. — For purposes of this section, the term "incremental cost of alternative electric energy" means, with respect to electric energy purchased from a qualifying cogenerator or qualifying small power producer, the cost to the electric utility of the electric energy which, but for the purchase from such cogenerator or small power producer, such utility would generate or purchase from another source.

(e) EXEMPTIONS. — (1) Not later than 1 year after the date of enactment of this Act and from time to time thereafter, the Commission shall, after consultation with representatives of State regulatory authorities, electric utilities, owners of cogeneration facilities and owners of small power production facilities, and after public notice and a reasonable opportunity for interested persons (including State and Federal agencies) to submit oral as well as written data, views, and arguments, prescribe rules under which qualifying cogeneration facilities and qualifying small power production facilities are exempted in whole or part from the Federal Power Act, from the Public Utility Holding Company Act, from State laws and regulations respecting the rates, or respecting the financial or organizational regulation, of electric utilities, or from any combination of the foregoing, if the Commission determines such exemption is necessary to encourage cogeneration and small power production.

(2) No qualifying small power production facility which has a power production capacity which, together with any other facilities located at the same site (as determined by the Commission), exceeds 30 megawatts may be exempted under rules under paragraph (1) from any provision of law or regulation referred to in paragraph (1), except that any qualifying small power production facility which produces electric energy solely by the use of biomass as a primary energy source may be exempted by the Commission under such rules from the Public Utility Holding Company Act and from State laws and regulations referred to in such paragraph (1).

(3) No qualifying small power production facility or qualifying cogeneration facility may be exempted under this subsection from —

> (A) any State law or regulation in effect in a State pursuant to subsection (f),
> (B) the provisions of section 210, 211, or 212 of the Federal Power Act or the necessary authorities for enforcement of any such provision under the Federal Power Act, or
> (C) any license or permit requirement under part I of the Federal Power Act, any provision under such Act related to such a license or permit requirement, or the necessary authorities for enforcement of any such requirement.

(f) IMPLEMENTATION OF RULES FOR QUALIFYING COGENERATION AND QUALIFYING SMALL POWER PRODUCTION FACILITIES. — (1) Beginning on or before the date one year after any rule is prescribed by the Commission under subsection (a) or revised under such subsection, each State regulatory authority shall, after notice and opportunity for public hearing, implement such rule (or revised rule) for each electric utility for which it has ratemaking authority.

SMALL QUALIFYING POWER PRODUCTION FACILITIES. — (1) Beginning on or before the date one year after any rule is prescribed by the Commission under subsection (a) or revised under such subsection, each State regulatory authority shall, after notice and opportunity for public hearing, implement such rule (or revised rule) for each electric utility for which it has ratemaking authority.

(2) Beginning on or before the date one year after any rule is prescribed by the Commission under subsection (a) or revised under such subsection, each nonregulated electric utility shall, after notice and opportunity for public hearing, implement such rule (or revised rule).

(g) JUDICIAL REVIEW AND ENFORCEMENT. — (1) Judicial review may be obtained respecting any proceeding conducted by a State regulatory authority or nonregulated electric utility for purposes of implementing any requirement of a rule under subsection (a) in the same manner, and under the same requirements, as judicial review may be obtained under section 123 in the case of a proceeding to which section 123 applies.

(2) Any person (including the Secretary) may bring an action against any electric utility, qualifying small power producer, or qualifying cogenerator to enforce any requirement established by a State regulatory authority or nonregulated electric utility pursuant to subsection (f). Any such action shall be brought only in the manner, and under the requirements, as provided under section 123 with respect to an action to which section 123 applies.

(h) COMMISSION ENFORCEMENT. — (1) For purposes of enforcement of any rule prescribed by the Commission under subsection (a) with respect to any operations of an electric utility, a qualifying cogeneration facility or a qualifying small power production facility which are subject to the jurisdiction of the Commission under part II of the Federal Power Act, such rule shall be treated as a rule under the Federal Power Act. Nothing in subsection (g) shall apply to so much of the operations of an electric utility, a qualifying cogeneration facility or a qualifying small power production facility as are subject to the jurisdiction of the Commission under part II of the Federal Power Act.

(2) (A) The Commission may enforce the requirements of subsection (f) against any State regulatory authority or nonregulated electric utility. For purposes of any such enforcement, the

requirements of subsection (f) (1) shall be treated as a rule enforceable under the Federal Power Act. For purposes of any such action, a State regulatory authority or nonregulated electric utility shall be treated as a person within the meaning of the Federal Power Act. No enforcement action may be brought by the Commission under this section other than —

(i) an action against the State regulatory authority or nonregulated electric utility for failure to comply with the requirements of subsection (f) or

(ii) an action under paragraph (1).

(B) Any electric utility, qualifying cogenerator, or qualifying small power producer may petition the Commission to enforce the requirements of subsection (f) as provided in subparagraph (A) of this paragraph. If the Commission does not initiate an enforcement action under subparagraph (A) against a State regulatory authority or nonregulated electric utility within 60 days following the date on which a petition is filed under this subparagraph with respect to such authority, the petitioner may bring an action in the appropriate United States district court to require such State regulatory authority or nonregulated electric utility to comply with such requirements, and such court may issue such injunctive or other relief as may be appropriate. The Commission may intervene as a matter of right in any such action.

(i) FEDERAL CONTRACTS. — No contract between a Federal agency and any electric utility for the sale of electric energy by such Federal agency for resale which is entered into after the date of the enactment of this Act may contain any provision which will have the effect of preventing the implementation of any rule under this section with respect to such utility. Any provision in any such contract which has such effect shall be null and void.

(j) DEFINITIONS. — For purposes of this section, the terms "small power production facility," "qualifying small power production facility," "qualifying small power producer," "primary energy source," "cogeneration facility," "qualifying cogeneration facility," and "qualifying cogenerator" have the respective meanings provided for such terms under section 3 (17) and (18) of the Federal Power Act.

Appendix 2

Federal Energy Regulatory Commission Regulations Implementing PURPA Sections 201 and 210

1. Part 292 of Subchapter K is amended by adding a new Subpart B to read as follows:

Subpart B — Qualifying Cogeneration and Small Power Production Facilities

Sec.
292.201   Scope.
292.202   Definitions.
292.203   General requirements for qualification.
292.204   Criteria for qualifying small power production facilities.
292.205   Criteria for qualifying cogeneration facilities.
292.206   Ownership criteria.
292.207   Procedures for obtaining qualifying status.

Authority: Public Utility Regulatory Policies Act of 1978, (16 U.S.C. 2601, *et seq.*), Energy Supply and Environmental Coordination Act, (15 U.S.C. 791 *et seq.*), Federal Power Act, as amended, (16 U.S.C. 792, *et seq.*), Department of Energy Organization Act, (42 U.S.C. 7101 *et seq.*), E.O. 12009, 42 FR 46267, Natural Gas Policy Act of 1978, (15 U.S.C. 3301, *et seq.*)

Subpart B — Qualifying Cogeneration and Small Power Production Facilities

§ 292.201   Scope.

This subpart applies to the criteria for and manner of becoming a qualifying small power production facility and a qualifying cogeneration facility under sections 3(17)(C) and 3(18)(B), respectively, of the Federal Power Act, as amended by section 201 of the Public Utility Regulatory Policies Act of 1978 (PURPA).

§ 292.202   Definitions.

For purposes of this subpart:

(a) "Biomass" means any organic material not derived from fossil fuels;

(b) "Waste" means by-product materials other than biomass;

(c) "Cogeneration facility" means equipment used to produce electric energy and forms of useful thermal energy (such as heat or steam), used for industrial, commercial, heating, or cooling purposes, through the sequential use of energy;

(d) "Topping-cycle cogeneration facility" means a cogeneration facility in which the energy input to the facility is first used to produce useful power output, and the reject heat from power production is then used to provide useful thermal energy;

(e) "Bottoming-cycle cogeneration facility" means a cogeneration facility in which the energy input to the system is first applied to a useful thermal energy process, and the reject heat emerging from the process is then used for power production;

(f) "Supplementary firing" means an energy input to the cogeneration facility used only in the thermal process of a topping-cycle cogeneration facility, or only in the electric generating process of a bottoming-cycle cogeneration facility;

(g) "Useful power output" of a cogeneration facility means the electric or mechanical energy made available for use, exclusive of any such energy used in the power production process;

(h) "Useful thermal energy output" of a topping-cycle cogeneration facility means the thermal energy made available for use in any industrial or commercial process, or used in any heating or cooling application;

(i) "Total energy output" of a topping-cycle cogeneration facility is the sum of the useful power output and useful thermal energy output;

(j) "Total energy input" means the total energy of all forms supplied from external sources other than supplementary firing to the facility;

(k) "Natural gas" means either natural gas unmixed, or any mixture of natural gas and artificial gas;

(l) "Oil" means crude oil, residual fuel oil, natural gas liquids, or any refined petroleum products; and

(m) Energy input in the case of energy in the form of natural gas or oil is to be measured by the lower heating value of the natural gas or oil.

§ 292.203   General requirements for qualification.

(a) *Small power production facilities.* A small power production facility is a qualifying facility if it:
(1) Meets the maximum size criteria specified in § 292.204(a);
(2) Meets the fuel use criteria specified in § 292.204(b); and
(3) Meets the ownership criteria specified in § 292.206.
(b) *Cogeneration facilities.* (1) Unless excluded under paragraph (c), a cogeneration facility is a qualifying facility if it:
(i) Meets any applicable operating and efficiency standards specified in § 292.205(a) and (b); and
(ii) Meets the ownership criteria specified in § 292.206
(2) For purposes of qualification of a cogeneration facility for exemption from incremental pricing, a cogeneration facility must qualify under § 292.205(c).
(c) *Interim exclusion.* (1) Pending further Commission action, any cogeneration facility which is a new diesel cogeneration facility may not be a qualifying facility.
(2) A new diesel cogeneration facility is a cogeneration facility:
(i) Which derives its useful power output from a diesel engine, and
(ii) The installation of which began on or after March 13, 1980.

§ 292.204   Criteria for qualifying small power production facilities.

(a) *Size of the facility* —(1) *Maximum size.* The power production capacity of the facility for which qualification is sought, together with the capacity of any other facilities which use the same energy resource, are owned by the same person, and are located at the same site, may not exceed 80 megawatts.
(2) *Method of calculation.* (i) For purposes of this paragraph, facilities are considered to be located at the same site as the facility for which qualification is sought if they are located within one mile of the facility for which qualification is sought and, for hydroelectric facilities, if they use water from the same impoundment for power generation.
(ii) For purposes of making the determination in clause (i), the distance between facilities shall be measured from the electrical generating equipment of a facility.

(3) *Waiver.* The Commission may modify the application of subparagraph (2) for good cause.

(b) *Fuel use.* (1)(i) The primary energy source of the facility must be biomass, waste, renewable resources, or any combination thereof, and more than 50 percent of the total energy input must be from these sources.

(ii) Any primary energy source which, on the basis of its energy content, is 50 percent or more biomass shall be considered biomass.

(2) Use of oil, natural gas, and coal by a facility may not, in the aggregate, exceed 25 percent of the total energy input of the facility during any calendar year period.

§ 292.205   Criteria for qualifying cogeneration facilities.

(a) *Operating and efficiency standards for topping-cycle facilities* —(1) *Operating standard.* For any topping-cycle cogeneration facility, the useful thermal energy output of the facility must, during any calendar year period, be no less than 5 percent of the total energy output.

(2) *Efficiency standard.* (i) For any topping-cycle cogeneration facility for which any of the energy input is natural gas or oil, and the installation of which began on or after March 13, 1980, the useful power output of the facility plus one-half the useful thermal energy output, during any calendar year period, must:

(A) Subject to paragraph (a)(2)(i)(B) of this section be no less than 42.5 percent of the total energy input of natural gas and oil to the facility; or

(B) If the useful thermal energy output is less than 15 percent of the total energy output of the facility, be no less than 45 percent of the total energy input of natural gas and oil to the facility.

(ii) For any topping-cycle cogeneration facility not subject to paragraph (a)(2)(i) of this section there is no efficiency standard.

(b) *Efficiency standards for bottoming-cycle facilities.* (1) For any bottoming-cycle cogeneration facility for which any of the energy input as supplementary firing is natural gas or oil, and the installation of which began on or after March 13, 1980, the useful power output of the facility must, during any calendar year period, be no less than 45 percent of the energy input of natural gas and oil for supplementary firing.

(2) For any bottoming-cycle cogeneration facility not covered by subparagraph (1) of this paragraph, there is no efficiency standard.

(c) *Exemption from incremental pricing.* (1) Natural gas used in any topping-cycle cogeneration facility is eligible for an exemption from incremental pricing under Title II of the Natural Gas Policy Act of 1978 (NGPA) and Part 282 of the Commission's rules if:

(i) The facility meets the operating and efficiency standards under paragraphs (a)(1) and (2)(i) of this section and is a qualifying facility under § 292.203(b)(1); or

(ii) The facility is a qualifying facility under Subpart E of this part.

(2) Natural gas used in any bottoming-cycle cogeneration facility, not subject to an exemption from incremental pricing under Subpart E of this part, is eligible for an exemption under Title II of the NGPA and Part 282 of the Commission's rules to the extent that reject heat emerging from the useful thermal energy process is made available for use for power production.

(3) Nothing in this subpart affects any exemption provided under Subpart E of this part.

(4) Natural gas used for supplementary firing in any cogeneration facility is not eligible under this part for exemption from incremental pricing.

(d) *Waiver.* The Commission may waive any of the requirements of paragraphs (a), (b) and (c) of this section upon a showing that the facility will produce significant energy savings.

§ 292.206   Ownership criteria.

(a) *General rule.* A cogeneration facility or small power production facility may not be owned by a person primarily engaged in the generation or sale of electric power (other than electric power solely from cogeneration facilities or small power production facilities).

(b) *Ownership test.* For purposes of this section, a cogeneration or small power production facility shall be considered to be owned by a person primarily engaged in the generation or sale of electric power, if more than 50 percent of the equity interest in the facility is held by an electric utility or utilities, or by a public utility holding company, or companies, or any combination thereof. If a wholly or partially owned subsidiary of an electric utility or public utility holding company has an ownership interest of a facility, the subsidiary's ownership interest shall be considered as ownership by an electric utility or public utility holding company.

§ 292.207   Procedures for obtaining qualifying status.

(a) *Qualification.* (1) A small power production facility or cogeneration facility which meets the criteria for qualification set forth in § 292.203 is a qualifying facility.

(2) The owner or operator of any facility qualifying under this paragraph shall furnish notice to the Commission providing the information set forth in paragraph (b)(2)(i) through (iv) of this section.

(b) *Optional procedure* —(1) *Application for Commission certification.* Pursuant to the provisions of this paragraph, the owner or operator of the facility may file with this Commission an application for Commission certification that the facility is a qualifying facility.

(2) *General contents of application.* The application shall contain the following information:

(i) The name and address of the applicant and location of the facility;

(ii) A brief description of the facility, including a statement indicating whether such facility is a small power production facility or a cogeneration facility;

(iii) The primary energy source used or to be used by the facility;

(iv) The power production capacity of the facility; and

(v) The percentage of ownership by any electricy utility or by any public utility holding company, or by any person owned by either.

(3) *Additional application requirements for small power production facilities.* An application by a small power producer for Commission certification shall contain the following additional information:

(i) The location of the facility in relation to any other small power production facilities located within one mile of the facility, owned by the applicant which use the same energy source; and

(ii) Information identifying any planned usage of natural gas, oil or coal.

(4) *Additional application requirements for cogeneration facilities.* An application by a cogenerator for Commission certification shall contain the following additional information;

(i) A description of the cogeneration system, including whether the facility is a topping or bottoming cycle and sufficient information to determine that any applicable requirements under § 292.205 will be met; and

(ii) The date installation of the facility began or will begin.

(5) *Commission action.* Within 90 days of the filing of an application, the Commission shall issue an order granting or denying the application, tolling the time for issuance of an order, or setting the matter for hearing. Any order denying certification shall identify the specific requirements which were not met. If no order is issued within 90 days of the filing of the complete application, it shall be deemed to have been granted.

(c) *Notice requirements for facilities of 500 kw or more.* An
electric utility is not required to purchase electric energy from a
facility with a design capacity of 500 kw or more until 90 days after
the facility notifies the utility that it is a qualifying facility, or 90
days after the facility has applied to the Commission under
paragraph (b) of this section.

(d) *Revocation of qualifying status.* (1) The Commission may
revoke the qualifying status of a qualifying facility which has been
certified under this section if such facility fails to comply with any
of the statements contained in its application for Commission
certification.

(2) Prior to undertaking any substantial alteration or modification
of a qualifying facility which has been certified under this section, a
small power producer or cogenerator may apply to the Commission
for a determination that the proposed alteration or modification will
not result in a revocation of qualifying status.

PART 292—REGULATIONS UNDER SECTIONS 201 AND 210 OF THE PUBLIC UTILITY REGULATORY POLICIES ACT OF 1978 WITH REGARD TO SMALL POWER PRODUCTION AND COGENERATION.

Subpart A—General Provisions

Subpart B—[Reserved]

Subpart C—Arrangements Between Electric Utilities and Qualifying Cogeneration and Small Power Production Facilities Under Section 210 of the Public Utility Regulatory Policies Act of 1978

Subpart D—Implementation

292.401   Implementation by State Regulatory Authorities and
     Nonregulated Utilities.
292.402   Implementation of Certain Reporting Requirements.
292.403   Waivers.
*   *   *   *   *

Subpart F—Exemption of Qualifying Small Power Production Facilities and Cogeneration Facilities from Certain Federal and State Laws and Regulations

292.601   Exemption of Qualifying Facilities from the Federal Power
     Act.
292.602   Exemption of Qualifying Facilities from the Public Utility
   Holding Company Act and Certain State Law and Regulation.
   Authority: This part issued under the Public Utility Regulatory
Policies Act of 1978, 16 U.S.C. § 2601 *et seq.*, Energy Supply and
Environmental Coordination Act. 15, U.S.C. § 791 *et seq.* Federal
Power Act , 16 U.S.C. § 792 *et seq.* Department of Energy
Organization Act, 42 U.S.C. § 7101 *et seq.*, E.O. 12009, 42 FR
46267.

Subpart A—General Provisions

§ 292.101   Definitions.

   (a) *General rule.* Terms defined in the Public Utility Regulatory
Policies Act of 1978 (PURPA) shall have the same meaning for
purposes of this part as they have under PURPA, unless further
defined in this part.
   (b) *Definitions.* The following definitions apply for purposes of
this part.
   (1) "Qualifying facility" means a cogeneration facility or a small
power production facility which is a qualifying facility under
Subpart B of this part of the Commission's regulations.
   (2) "Purchase" means the purchase of electric energy or capacity
or both from a qualifying facility by an electric utility.
   (3) "Sale" means the sale of electric energy or capacity or both
by an electric utility to a qualifying facility.
   (4) "System emergency" means a condition on a utility's system
which is likely to result in imminent significant disruption of service
to customers or is imminently likely to endanger life or property.
   (5) "Rate" means by price, rate, charge, or classification made,
demanded, observed or received with respect to the sale or
purchase of electric energy or capacity, or any rule, regulation, or

practice respecting any such rate, charge, or classification, and any contract pertaining to the sale or purchase of electric energy or capacity.

(6) "Avoided costs" means the incremental costs to an electric utility of electric energy or capacity or both which, but for the purchase from the qualifying facility or qualifying facilities, such utility would generate itself or purchase from another source.

(7) "Interconnection costs" means the reasonable costs of connection, switching, metering, transmission, distribution, safety provisions and administrative costs incurred by the electric utility directly related to the installation and maintenance of the physical facilities to permit interconnected operations with a qualifying facility, to the extent such costs are in excess of the corresponding costs which the electric utility would have incurred if it had not engaged in interconnected operations, but instead generated an equivalent amount of electric energy itself or purchased an equivalent amount of electric energy or capacity from other sources. Interconnection costs do not include any costs included in the calculation of avoided costs.

(8) "Supplementary power" means electric energy or capacity supplied by an electric utility, regularly used by a qualifying facility in addition to that which the facility generates itself.

(9) "Back-up power" means electric energy or capacity supplied by an electric utility to replace energy ordinarily generated by a facility's own generation equipment during an unscheduled outage of the facility.

(10) "Interruptible power" means electric energy or capacity supplied by an electric utility subject to interruption by the electric utility under specified conditions.

(11) "Maintenance power" means electric energy or capacity supplied by an electric utility during scheduled outages of the qualifying facility.

Subpart B—[Reserved]

Subpart C—Arrangements Between Electric Utilities and Qualifying Cogeneration and Small Power Production Facilities Under Section 210 of the Public Utility Regulatory Policies Act of 1978

§ 292.301 Scope.

(a) *Applicability.* This subpart applies to the regulation of sales and purchases between qualifying facilities and electric utilities.

(b) *Negotiated rates or terms.* Nothing in this subpart:

(1) Limits the authority of any electric utility or any qualifying facility to agree to a rate for any purchase, or terms or conditions relating to any purchase, which differ from the rate or terms or conditions which would otherwise be required by this subpart; or

(2) Affects the validity of any contract entered into between a qualifying facility and an electric utility for any purchase.

§ 292.302   Availability of electric utility system cost data.

(a) *Applicability.* (1) Except as provided in paragraph (a)(2) of this section, paragraph (b) applies to each electric utility, in any calendar year, if the total sales of electric energy by such utility for purposes other than resale exceeded 500 million kilowatt-hours during any calendar year beginning after December 31, 1975, and before the immediately preceding calendar year.

(2) Each utility having total sales of electric energy for purposes other than resale of less than one billion kilowatt-hours during any calendar year beginning after December 31, 1975, and before the immediately preceding year, shall not be subject to the provisions of this section until May 31, 1982.

(b) *General rule.* To make available data from which avoided costs may be derived, not later than November 1, 1980, May 31, 1982, and not less often than every two years thereafter, each regulated electric utility described in paragraph (a) of this section shall provide to its State regulatory authority, and shall maintain for public inspection, and each nonregulated electric utility described in paragraph (a) of this section shall maintain for public inspection, the following data:

(1) The estimated avoided cost on the electric utility's system, solely with respect to the energy component, for various levels of purchases from qualifying facilities. Such levels of purchases shall be stated in blocks of not more than 100 megawatts for systems with peak demand of 1000 megawatts or more, and in blocks equivalent to not more than 10 percent of the system peak demand for systems of less than 1000 megawatts. The avoided costs shall be stated on a cents per kilowatt-hour basis, during daily and seasonal peak and off-peak periods, by year, for the current calendar year and each of the next 5 years;

(2) The electric utility's plan for the addition of capacity by amount and type, for purchases of firm energy and capacity, and for capacity retirements for each year during the succeeding 10 years; and

(3) The estimated capacity costs at completion of the planned capacity additions and planned capacity firm purchases, on the basis of dollars per kilowatt, and the associated energy costs of each unit, expressed in cents per kilowatt hour. These costs shall be expressed in terms of individual generating units and of individual planned firm purchases.

(c) *Special rule for small electric utilities.*

(1) Each electric utility (other than any electric utility to which paragraph (b) of this section applies) shall, upon request:

(i) Provide comparable data to that required under paragraph (b) of this section to enable qualifying facilities to estimate the electric utility's avoided costs for periods described in paragraph (b) of this section; or

(ii) With regard to an electric utility which is legally obligated to obtain all its requirements for electric energy and capacity from another electric utility, provide the data of its supplying utility and the rates at which it currently purchases such energy and capacity.

(2) If any such electric utility fails to provide such information on request, the qualifying facility may apply to the State regulatory authority (which has ratemaking authority over the electric utility) or the Commission for an order requiring that the information be provided.

(d) *Substitution of alternative method.* (1) After public notice in the area served by the electric utility, and after opportunity for public comment, any State regulatory authority may require with respect to any electric utility over which it has ratemaking authority), or any non-regulated electric utility may provide data different than those which are otherwise required by this section if it determines that avoided costs can be derived from such data.

(2) Any State regulatory authority with respect to any electric utility over which it has ratemaking authority) or nonregulated utility which requires such different data shall notify the Commission within 30 days of making such determination.

(e) *State Review.* (1) Any data submitted by an electric utility under this section shall be subject to review by the State regulatory authority which has ratemaking authority over such electric utility.

(2) In any such review, the electric utility has the burden of coming forward with justification for its data.

§ 292.303   Electric utility obligations under this subpart.

(a) *Obligation to purchase from qualifying facilities.* Each electric utility shall purchase, in accordance with § 292.304, any energy and capacity which is made available from a qualifying facility:

(1) Directly to the electric utility; or

(2) Indirectly to the electric utility in accordance with paragraph (d) of this section.

(b) *Obligation to sell to qualifying facilities.* Each electric utility shall sell to any qualifying facility, in accordance with § 292.305, any energy and capacity requested by the qualifying facility.

(c) *Obligation to interconnect.* (1) subject to paragraph (c)(2) of this section, any electric utility shall make such interconnections with any qualifying facility as may be necessary to accomplish purchases or sales under this subpart. The obligation to pay for any interconnection costs shall be determined in accordance with § 292.306.

(2) No electric utility is required to interconnect with any qualifying facility if, solely by reason of purchases or sales over the interconnection, the electric utility would become subject to regulation as a public utility under Part II of the Federal Power Act.

(d) *Transmission to other electric utilities.* If a qualifying facility agrees, an electric utility which would otherwise be obligated to purchase energy or capacity from such qualifying facility may transmit the energy or capacity to any other electric utility. Any electric utility to which such energy or capacity is transmitted shall purchase such energy or capacity under this subpart as if the qualifying facility were supplying energy or capacity directly to such electric utility. The rate for purchase by the electric utility to which such energy is transmitted shall be adjusted up or down to reflect line losses pursuant to § 292.304(e)(4) and shall not include any charges for transmission.

(e) *Parallel operation.* Each electric utility shall offer to operate in parallel with a qualifying facility, provided that the qualifying facility complies with any applicable standards established in accordance with § 292.308.

§ 292.304   Rates for purchases.

(a) *Rates for purchases.* (1) Rates for purchases shall:

(i) Be just and reasonable to the electric consumer of the electric utility and in the public interest; and

(ii) Not discriminate against qualifying cogeneration and small power production facilities.

(2) Nothing in this subpart requires any electric utility to pay more than the avoided costs for purchases.

(b) *Relationship to avoided costs.* (1) For purposes of this paragraph, "new capacity" means any purchase from capacity of a qualifying facility, construction of which was commenced on or after November 9, 1978.

(2) Subject to paragraph (b)(3) of this section, a rate for purchases satisfies the requirements of paragraph (a) of this section if the rate equals the avoided costs determined after consideration of the factors set forth in paragraph (e) of this section.

(3) A rate for purchases (other than from new capacity) may be less than the avoided cost if the State regulatory authority (with respect to any electric utility over which it has ratemaking authority) or the nonregulated electric utility determines that a lower rate is consistent with paragraph (a) of this section, and is sufficient to encourage cogeneration and small power production.

(4) Rates for purchases from new capacity shall be in accordance with paragraph (b)(2) of this section, regardless of whether the electric utility making such purchases is simultaneously making sales to the qualifying facility.

(5) In the case in which the rates for purchases are based upon estimates of avoided costs over the specific term of the contract or other legally enforceable obligation, the rates for such purchases do not violate this subpart if the rates for such purchases differ from avoided costs at the time of delivery.

(c) *Standard rates for purchases.* (1) There shall be put into effect (with respect to each electric utility) standard rates for purchases from qualifying facilities with a design capacity of 100 kilowatts or less.

(2) There may be put into effect standard rates for purchases from qualifying facilities with a design capacity of more than 100 kilowatts.

(3) The standard rates for purchases under this paragraph:

(i) Shall be consistent with paragraphs (a) and (e) of this section; and

(ii) May differentiate among qualifying facilities using various technologies on the basis of the supply characteristics of the different technologies.

(d) *Purchases "as available" or pursuant to a legally enforceable obligation.* Each qualifying facility shall have the option either:

(1) To provide energy as the qualifying facility determines such energy to be available for such purchases, in which case the rates for such purchases shall be based on the purchasing utility's avoided costs calculated at the time of delivery; or

(2) To provide energy or capacity pursuant to a legally enforceable obligation for the delivery of energy or capacity over a specified term, in which case the rates for such purchases shall, at

the option of the qualifying facility exercised prior to the beginning
of the specified term, be based on either:

(i) The avoided costs calculated at the time of delivery; or

(ii) The avoided costs calculated at the time the obligation is
incurred.

(e) *Factors affecting rates for purchases.* In determining avoided
costs, the following factors shall, to the extent practicable, be taken
into account:

(1) The data provided pursuant to § 292.302(b), (c), or (d),
including State review of any such data;

(2) The availability of capacity or energy from a qualifying facility
during the system daily and seasonal peak periods, including:

(i) The ability of the utility to dispatch the qualifying facility;

(ii) The expected or demonstrated reliability of the qualifying
facility;

(iii) The terms of any contract or other legally enforceable
obligation, including the duration of the obligation, termination
notice requirement and sanctions for non-compliance;

(iv) The extent to which scheduled outages of the qualifying
facility can be usefully coordinated with scheduled outages of the
utility's facilities;

(v) The usefulness of energy and capacity supplied from a
qualifying facility during system emergencies, including its ability to
separate its load from its generation;

(vi) The individual and aggregate value of energy and capacity
from qualifying facilities on the electric utility's system; and

(vii) The smaller capacity increments and the shorter lead times
available with additions of capacity from qualifying facilities; and

(3) The relationship of the availability of energy or capacity from
the qualifying facility as derived in paragraph (e)(2) of this section,
to the ability of the electric utility to avoid costs, including the
deferral of capacity additions and the reduction of fossil fuel use;
and

(4) The costs or savings resulting from variations in line losses
from those that would have existed in the absence of purchases
from a qualifying facility, if the purchasing electric utility generated
an equivalent amount of energy itself or purchased an equivalent
amount of electric energy or capacity.

(f) *Periods during which purchases not required.*

(1) Any electric utility which gives notice pursuant to paragraph
(f)(2) of this section will not be required to purchase electric energy
or capacity during any period during which, due to operational
circumstances, purchases from qualifying facilities will result in

costs greater than those which the utility would incur if it did not make such purchases, but instead generated an equivalent amount of energy itself.

(2) Any electric utility seeking to invoke paragraph (f)(1) of this section must notify, in accordance with applicable State law or regulation, each affected qualifying facility in time for the qualifying facility to cease the delivery of energy or capacity to the electric utility.

(3) Any electric utility which fails to comply with the provisions of paragraph (f)(2) of this section will be required to pay the same rate for such purchase of energy or capacity as would be required had the period described in paragraph (f)(1) of this section not occurred.

(4) A claim by an electric utility that such a period has occurred or will occur is subject to such verification by its State regulatory authority as the State regulatory authority determines necessary or appropriate, either before or after the occurrence.

§ 292.305   Rates for sales.

(a) *General rules.* (1) Rates for sales:

(i) Shall be just and reasonable and in the public interest; and

(ii) Shall not discriminate against any qualifying facility in comparison to rates for sales to other customers served by the electric utility.

(2) Rates for sales which are based on accurate data and consistent systemwide costing principles shall not be considered to discriminate against any qualifying facility to the extent that such rates apply to the utility's other customers with similar load or other cost-related characteristics.

(b) *Additional Services to be Provided to Qualifying Facilities.* (1) Upon request of a qualifying facility, each electric utility shall provide:

(i) Supplementary power;

(ii) Back-up power;

(iii) Maintenance power; and

(iv) Interruptible power.

(2) The State regulatory authority (with respect to any electric utility over which it has ratemaking authority) and the Commission (with respect to any nonregulated electric utility) may waive any requirement of paragraph (b)(1) of this section if, after notice in the area served by the electric utility and after opportunity for public comment, the electric utility demonstrates and the State regulatory authority or the Commission, as the case may be, finds that compliance with such requirement will:

(i) Impair the electric utility's ability to render adequate service to its customers; or

(ii) Place an undue burden on the electric utility.

(c) *Rates for sales of back-up and maintenance power.* The rate for sales of back-up power or maintenance power:

(1) shall not be based upon an assumption (unless supported by factual data) that forced outages or other reductions in electric output by all qualifying facilities on an electric utility's system will occur simultaneously, or during the system peak, or both; and

(2) shall take into account the extent to which scheduled outages of the qualifying facilities can be usefully coordinated with scheduled outages of the utility's facilities.

§ 292.306   Interconnection costs.

(a) *Obligation to pay.* Each qualifying facility shall be obligated to pay any interconnection costs which the State regulatory authority (with respect to any electric utility over which it has ratemaking authority) or nonregulated electric utility may assess against the qualifying facility on a nondiscriminatory basis with respect to other customers with similar load characteristics.

(b) *Reimbursement of interconnection costs.* Each State regulatory authority (with respect to any electric utility over which it has ratemaking authority) and nonregulated utility shall determine the manner for payments of interconnection costs, which may include reimbursement over a reasonable period of time.

§ 292.307   System emergencies.

(a) *Qualifying facility obligation to provide power during system emergencies.* A qualifying facility shall be required to provide energy or capacity to an electric utility during a system emergency only to the extent:

(1) Provided by agreement between such qualifying facility and electric utility; or

(2) Ordered under section 202(c) of the Federal Power Act.

(b) *Discontinuance of purchases and sales during system emergencies.* During any system emergency, an electric utility may discontinue:

(1) Purchases from a qualifying facility if such purchases would contribute to such emergency; and

(2) Sales to a qualifying facility, provided that such discontinuance is on a nondiscriminatory basis.

§ 292.308   Standards for operating reliability.

Any State regulatory authority (with respect to any electric utility over which it has ratemaking authority) or nonregulated electric utility may establish reasonable standards to ensure system safety and reliability of interconnected operations. Such standards may be recommended by any electric utility, any qualifying facility, or any other person. If any State regulatory authority (with respect to any electric utility over which it has ratemaking authority) or nonregulated electric utility establishes such standards, it shall specify the need for such standards on the basis of system safety and reliability.

Subpart D — Implementation

§ 292.401   Implementation by State regulatory authorities and nonregulated electric utilities.

(a) *State regulatory authorities.* Not later than one year after these rules take effect, each State regulatory authority shall, after notice and an opportunity for public hearing, commence implementation of Subpart C (other than § 292.302 thereof). Such implementation may consist of the issuance of regulations, an undertaking to resolve disputes between qualifying facilities and electric utilities arising under Subpart C, or any other action reasonably designed to implement such subpart (other than § 292.302 thereof).

(b) *Nonregulated electric utilities.* Not later than one year after these rules take effect, each nonregulated electric utility shall, after notice and an opportunity for public hearing, commence implementation of Subpart C (other than § 292.302 thereof). Such implementation may consist of the issuance of regulations, an undertaking to comply with Subpart C, or any other action reasonably designed to implement such subpart (other than § 292.302 thereof).

(c) *Reporting requirement.* Not later than one year after these rules take effect, each State regulatory authority and nonregulated electric utility shall file with the Commission a report describing the manner in which it will implement Subpart C (other than § 292.302 thereof).

§ 292.402   Implementation of certain reporting requirements.

Any electric utility which fails to comply with the requirements of § 292.302(b) shall be subject to the same penalties to which it may be subjected for failure to comply with the requirements of the Commission's regulations issued under section 133 of PURPA.

§ 292.403   Waivers.

(a) *State regulatory authority and nonregulated electric utility waivers.* Any State regulatory authority (with respect to any electric utility over which it has ratemaking authority) or nonregulated electric utility may, after public notice in the area served by the electric utility, apply for a waiver from the application of any of the requirements of Subpart C (other than § 292.302 thereof).

(b) *Commission action.* The Commission will grant such a waiver only if an applicant under paragraph (a) of this section demonstrates that compliance with any of the requirements of Subpart C is not necessary to encourage cogeneration and small power production and is not otherwise required under section 210 of PURPA.

Subpart F—Exemption of Qualifying Small Power Production Facilities and Cogeneration Facilities from Certain Federal and State Laws and Regulations
§ 292.601   Exemption to qualifying facilities from the Federal Power Act.

(a) *Applicability.* This section applies to:

(1) qualifying cogeneration facilities; and

(2) qualifying small power production facilities which have a power production capacity which does not exceed 30 megawatts.

(b) *General rule.* Any qualifying facility described in paragraph (a) shall be exempt from all sections of the Federal Power Act, except:

(1) Sections 1–30;

(2) Sections 202(c), 210, 211, and 212;

(3) Sections 305(c); and

(4) Any necessary enforcement provision of Part III with regard to the sections listed in paragraphs (b) (1), (2) and (3) of this section.

§ 292.602   Exemption to qualifying facilities from the Public Utility Holding Company Act and certain State law and regulation.

(a) *Applicability.* This section applies to any qualifying facility described in § 292.601(a), and to any qualifying small power production facility with a power production capacity over 30 megawatts if such facility produces electric energy solely by the use of biomass as a primary energy source.

(b) *Exemption from the Public Utility Holding Company Act of*

1935. A qualifying facility described in paragraph (a) shall not be considered to be an "electric utility company" as defined in section 2(a)(3) of the Public Utility Holding Company Act of 1935, 15 U.S.C. 79b(a)(3).

(c) *Exemption from certain State law and regulation.*

(1) Any qualifying facility shall be exempted (except as provided in paragraph (c)(2)) of this section from State law or regulation respecting:

(i) The rates of electric utilities; and

(ii) The financial and organizational regulation of electric utilities.

(2) A qualifying facility may not be exempted from State law and regulation implementing Subpart C.

(3) Upon request of a State regulatory authority or nonregulated electric utility, the Commission may consider a limitation on the exemptions specified in subparagraph (1).

(4) Upon request of any person, the Commission may determine whether a qualifying facility is exempt from a particular State law or regulation.

[FR Doc. 80–5720 Filed 2–22–80; 8:45 am]

BILLING CODE 6450–85–M

Appendix 3
Electricity Generated by Oil

GEOGRAPHICAL LOCATION	ELECTRICITY GENERATED BY OIL (10^9 kwh)	TOTAL ELECTRICITY GENERATED (10^9 kwh)	ELECTRICITY GENERATED BY OIL (%)
Northeast	3.91	6.66	59
Maine	0.10	0.58	17
N.H.	0.18	0.47	38
Vt.		0.38	
Mass.	2.62	3.05	86
R.I.	0.04	0.09	44
Conn.	0.96	2.10	46
Middle Atlantic	5.66	24.35	23
N.Y.	3.57	10.13	35
N.J.	0.88	3.33	26
Pa.	1.20	10.89	11
East North Central	1.44	36.09	4
Ohio	0.12	9.58	1
Ind.	0.04	6.47	1
Ill.	0.83	9.95	8
Mich.	0.43	6.77	6
Wis.	0.03	3.33	1
West North Central	0.15	16.40	1
Minn.	0.02	2.49	1
Iowa	0.01	1.82	1
Mo.	0.04	5.23	1
N.Dak.		1.49	
S.Dak.		0.94	
Nebr.		1.66	
Kans.	0.08	2.77	3
South Atlantic	7.58	39.37	19
Del.	0.37	0.67	55
Md.	0.47	3.08	15
D.C.	0.10	0.10	100
Va.	0.98	3.27	30
W.Va.	0.03	5.70	*

SOURCE: Energy Information Administration, *Electric Power Monthly*, August, 1980.
*Indicates less than 1%.

GEOGRAPHICAL LOCATION	ELECTRICITY GENERATED BY OIL (10⁹ kwh)	TOTAL ELECTRICITY GENERATED (10⁹ kwh)	ELECTRICITY GENERATED BY OIL (%)
N.C.	0.05	6.26	1
S.C.	0.41	3.89	11
Ga.	0.15	6.49	2
Fla.	5.03	9.91	51
East South Central	0.47	20.49	2
Ky.	0.01	5.23	*
Tenn.	0.02	5.34	*
Ala.	0.01	7.77	*
Miss.	0.44	2.15	20
West South Central	1.19	33.56	4
Ark.	0.44	2.41	18
La.	0.64	4.82	13
Okla.		4.85	
Tex.	0.11	21.48	*
Mountain	0.34	14.70	2
Mont.		1.30	
Idaho		0.67	
Wyo.		1.96	
Colo.	0.02	2.22	1
N.Mex.	0.01	2.34	*
Ariz.	0.14	3.84	4
Utah		1.09	
Nev.	0.16	1.28	13
Pacific Contiguous	3.54	23.06	15
Wash.		6.36	
Oreg.		2.93	
Calif.	3.53	13.77	26
Pacific Noncontiguous	0.58	0.77	75
Alaska	0.03	0.22	14
Hawaii	0.55	0.55	100
U.S. Total	24.85	215.44	12

Appendix 4
Twenty-Five Most Expensive and Twenty-Five Least Expensive Service Territories

Most Expensive Service Territories

CITY AND STATE	NAME OF COMPANY	1,500 KWH* TOTAL COST ($)	AVERAGE COST (¢/kwh)	RANKING
New York, N.Y.	Consolidated Edison of New York	228.88	15.26	1
Lihue, Hawaii	Kauai Electric	226.68	15.11	2
Nantucket, Mass.	Nantucket Electric	209.03	13.94	3
Wailuku, Hawaii	Maui Electric	207.79	13.85	4
Middletown, N.Y.	Orange & Rockland Utilities	205.68	13.71	5
Honolulu, Hawaii	Hawaiian Electric	184.74	12.32	6
Hilo, Hawaii	Hawaii Electric Light	178.54	11.90	7
Long Island, N.Y.	Long Island Lighting	175.16	11.68	8
San Diego, Calif.	San Diego Gas & Electric	173.19	11.55	9
Ramsey, N.J.	Rockland Electric	162.69	10.85	10
Wilmington, Del.	Delmarva Power & Light	155.79	10.39	11
Bridgeport, Conn.	United Illuminating	155.55	10.37	12
Chincoteague, Va.	Delmarva Power & Light	153.60	10.24	13
Poughkeepsie, N.Y.	Central Hudson Gas & Electric	152.84	10.19	14
Fitchburg, Mass.	Fitchburg Gas & Electric Light	148.68	9.91	15
Manchester, N.H.	Public Service of New Hampshire	148.54	9.90	16
Boston, Mass.	Boston Edison	148.42	9.89	17
Philadelphia, Pa.	Philadelphia Electric	144.48	9.63	18
Manchester, Mass.	Manchester Electric	143.53	9.57	19
Dover, N.J.	Jersey Central Power & Light	142.78	9.52	20
Chicago, Ill.	Commonwealth Edison	142.76	9.52	21
Elkton, Md.	Conowingo Power	141.14	9.41	22
Hartford, Conn.	Hartford Electric Light	140.66	9.38	23
Brockton, Mass.	Eastern Edison	138.93	9.26	24
Cleveland, Ohio	Cleveland Electric Illuminating	138.60	9.24	25

*Total cost = 500 kwh per month (June 1982, July 1982, August 1982).

Least Expensive Service Territories

CITY AND STATE	NAME OF COMPANY	1,500 KWH* TOTAL COST ($)	AVERAGE COST (¢/kwh)	RANKING
Noxon, Mont.	Washington Water	30.00	2.00	218
Kalispell, Mont.	Pacific Power & Light	32.70	2.18	217
Spokane, Wash.	Washington Water Power	34.15	2.28	216
Yakima, Wash.	Pacific Power & Light	41.78	2.79	215
Bellevue, Wash.	Puget Sound Power & Light	42.76	2.85	214
LaGrande, Oreg.	CP National	44.07	2.94	213
Ontario, Oreg.	Idaho Power	45.48	3.03	212
Lewiston, Idaho	Washington Water Power	45.75	3.05	211
Portland, Oreg.	Pacific Power & Light	46.17	3.08	210
Cheyenne, Wyo.	Cheyenne Light, Fuel & Power	47.91	3.19	209
Portland, Oreg.	Portland General Electric	50.57	3.37	208
Billings, Mont.	Montana Power	51.42	3.43	207
Siskiyou County, Calif.	Pacific Power & Light	54.03	3.60	206
Casper, Wyo.	Pacific Power & Light	58.56	3.90	205
Franklin, N.C.	Nantahala Power & Light	63.30	4.22	204
Boise, Idaho	Idaho Power	64.60	4.31	203
Rexburg, Idaho	Utah Power & Light	70.82	4.72	202
Gravette, Ark.	Empire District Electric	72.51	4.83	201
Rapid City, S.Dak.	Black Hills Power & Light	73.25	4.88	200
Ashland, Ky.	Kentucky Power	75.12	5.01	199
Flint, Mich.	Consumer's Power	77.48	5.17	198
North Sioux City, S.Dak.	Idaho Public Service	78.21	5.21	197
Las Vegas, Nev.	Nevada Power	78.30	5.22	196
Kingsport, Tenn.	Kingsport Power	78.59	5.24	195
Marietta, Ohio	Monongahela Power	79.02	5.27	194

* Total cost = 500 kwh per month (June 1982, July 1982, August 1982).

Appendix 5
Cost-of-Service Data for Selected Utilities

TABLE A5–1

FERC Rules, Section 292.302, Implementing Section 210 of PURPA
The Estimated Avoided Cost on the Electric Utility's System,
Solely with Respect to the Energy Component

	1982	1983	1984	1985	1986	1987
Jan.	0.988	1.721	2.237	2.492	2.759	3.054
Feb.	1.015	1.768	2.298	2.560	2.834	3.137
Mar.	0.988	1.721	2.237	2.492	2.759	3.054
Apr.	0.947	1.650	2.145	2.390	2.646	2.929
May	0.938	1.634	2.124	2.366	2.619	2.899
June	0.822	1.432	1.862	2.074	2.296	2.542
July	0.816	1.421	1.847	2.058	2.278	2.522
Aug.	0.816	1.421	1.847	2.058	2.278	2.522
Sept.	0.872	1.519	1.975	2.200	2.435	2.696
Oct.	1.512	1.966	2.190	2.424	2.683	2.970
Nov.	1.519	1.975	2.200	2.435	2.696	2.984
Dec.	1.721	2.237	2.492	2.759	3.054	3.381
Average	1.081	1.706	2.127	2.358	2.611	2.890

NOTE: All figures are cents per kwh in nominal dollars. Figures shown are
BPA minimum energy costs with the following escalation rates assumed:
Oct. 1982, 74.2%; Oct. 1983, 30%; Oct. 1984, 11.4%; Oct. 1985, 10.7%;
Oct. 1986, 10.7%; Oct. 1987, 10.7% (from *1982 Revenue Needs Analysis
Data*). The above escalation rates were based on a computer model that used
information that was available in May 1982.

TABLE A5–2

FERC Rules, Section 292.302, Implementing Section 210 of PURPA
The Estimated Capacity Costs at Completion of the Planned
Capacity Additions . . . and the Associated Energy Costs
of Each Unit . . .

PLANT	CAPACITY COSTS ($/Kw)*	ENERGY COSTS (¢/kwh)†	ON-LINE
Centralia (coal, 105 Mw)	190	1.2	1983‡
South Fork Tolt (hydro, 15 Mw)	2,103	0.48	1985
High Ross (hydro, 254 Mw)	923	0.02	1986

*Capacity costs are expressed in the dollars of the on-line year.
†Energy costs are 1981 dollars.
‡Seattle City Light began accepting energy and capacity from Centralia in 1983. The project was built in 1972. Capacity costs of Centralia are in 1972 dollars.

TABLE A5–3

Houston Lighting and Power Company
Schedule 302 (b) (1)
Estimates of Avoided Energy Costs for Purchases
of 100 Mw
(¢/kwh)

YEAR	SUMMER		WINTER
	ON-PEAK	OFF-PEAK	ALL HOURS
1982	5.03	4.41	4.79
1983	5.91	4.98	5.39
1984	6.65	5.59	5.74
1985	12.51	9.77	10.74
1986	10.79	8.62	9.03
1987	10.85	8.80	10.08

NOTES: Energy costs are in current (as-spent) dollars. All costs are shown at the generation level.
A number of individual factors shall be considered and evaluated in determining the company's avoided costs. These factors are set forth in § 292.304 (e) of the regulations of the Federal Energy Regulatory Commission. The data provided in this schedule are merely estimates. They are not intended to represent rates for future purchases by the company from qualifying facilities.
Avoided cost estimates were also developed for purchases of 500 Mw and 1,000 Mw. These costs were slightly lower than the above.

TABLE A5–4

Houston Lighting and Power Company
Schedule 302 (b) (3)
Estimated Capacity and Energy Costs for Planned
Capacity Additions and Firm Purchases

YEAR OF COMMISSIONING	UNIT NAME	CAPACITY COST ($/Kw)*†	ENERGY COST (¢/Kwh)‡
	Planned Purchases§	22	4.50
1983	W. A. Parish 8	680	4.45
	Deepwater Lowside‖	. . .	9.11
1986	Limestone 1	1,153	3.94
1987	South Texas 1	1,840	1.09
	Limestone 2	1,153	4.15
1988	Malakoff 1	1,460	3.64
1989	South Texas 2	1,840	1.20
	Malakoff 2	1,460	3.89
1991	A 1	1,472	8.63

* In as-spent $/kw of gross capability.
† All capacity costs exclude AFUDC.
‡ At maturity date in as-spent dollars. Maturity occurs within the third year of operation.
§ For 1982 purchases in $/kw per year.
‖ A rehabilitation project.

Appendix 6

Source List: Equipment and Publications

Cogeneration Equipment

Agway Inc.
Box 4933
Syracuse, NY 13221
(315) 477-7061

Alturdyne
8050 Armour St.
San Diego, CA 92111
(619) 565-2131

Caterpillar Engine
Peterson Power Center
2828 Teagarden St.
San Leandro, CA 94577
(415) 895-8400

Cogenic Energy Systems
645 Fifth Ave.
New York, NY 10022
(212) 832-6767

Cummins Engine Co.
1000 Fifth St.
Columbus, IN 47202
(812) 372-7211

Dresser Industries
Waukesha Engine Division
1000 St. Paul Ave.
Waukesha, WI 53187
(414) 547-3311

Kohler Co.
Kohler, WI 53044
(414) 457-4441

Re-Energy Systems, Inc.
660 W. Baltimore Pike
Media, PA 19063
(215) 565-9779

White Engine Co.
101 Eleventh St., SE
Canton, OH 44707
(216) 454-5631

Hydro Equipment

Allis-Chalmers
Hydro Turbines Division
Box 712
York, PA 17405
(717) 792-3511

Associated Electric Co., Inc.
54 Second Ave.
Chicopee, MA 01020
(413) 781-1053

Barber Hydraulic Turbines
Div. of Marsh Engineering Ltd.
Barber Point, P.O. Box 340
Port Colborne, ON L3K 5W1
Canada
(416) 834-9303

Canyon Industries, Inc.
5346 Mosquito Lake Rd.
Deming, WA 98244
(206) 592-5552

Cornell Pump Co., Inc.
2323 S.E. Harvester Dr.
Portland, OR 97222
(503) 653-0330

Energy Research and
   Applications, Inc.
1820 Fourteenth St.
Santa Monica, CA 90404
(213) 452-4905

Essex Turbines Co., Inc.
Kettle Cove Industrial Park
Magnolia, MA 01930
(617) 525-2011

Hydro Watt Systems, Inc.
146 Siglun Rd.
Coos Bay, OR 97420
(503) 267-3559

McKay Water Power, Inc.
P.O. Box 221
West Lebanon, NH 03784
(603) 298-5122

New England Energy
   Development Systems,
   Inc. (NEEDS)
109 Main St.
Amherst, MA 01002
(413) 256-8466

New Found Power Co., Inc.
Box 576
Hope Valley, RI 02832
(401) 539-2335

The Schneider Engine Co.
Rt. 1, Box 81
Justin, TX 76247
(817) 430-0174

Small Hydro-Electrics
   Canada Ltd.
P.O. Box 54
Silverton, BC V0G 280
Canada
(604) 358-2406

Worthington Division
McGraw-Edison Co.
5310 Taneytown Pike
Taneytown, MD 21787
(301) 756-2602

Photovoltaic Modules and Arrays

Acurex Corp.
485 Clyde Ave.
Mountain View, CA 94042
(415) 964-3200

ARCO Solar Industries, Inc.
20554 Plummer St.
Chatsworth, CA 91311
(213) 700-7000

International Rectifier Inc.
233 Kansas St.
El Segundo, CA 90245
(213) 322-3331

Photon Power Inc.
13 Founders Rd.
El Paso, TX 79906
(915) 779-7774

Solarex Corp.
1335 Piccard Dr.
Rockville, MD 20850
(301) 948-0202

Solar Power Corp.
20 Cabot Rd.
Woburn, MA 01801
(617) 935-4600

Solarwest Electric
232 Anacapa St.
Santa Barbara, CA 93101
(805) 963-9667

Solec International Inc.
12533 Chadron Ave.
Hawthorne, CA 90250
(213) 970-0065

Utility Interface Hardware

Aeolian Kinetics, Inc.
P.O. Box 100
Providence, RI 02901
(401) 421-5033

Beckwith Electric Co.
11811 Sixty-second St., North
Largo, FL 33543
(813) 535-3408

Best Energy Systems for
Tomorrow, Inc.
P.O. Box 280
Necedah, WI 54646
(608) 565-7200

Carter Motor Co.
2711 W. George St.
Chicago, IL 60618
(312) 588-7700

Climet Instruments Co.
1320 W. Colton Ave.
Redlands, CA 92373
(714) 793-2788

Danforth Co.
500 Riverside Industrial Pkwy.
Portland, ME 04103
(207) 797-2791

Davis Instrument Manufacturing
Co., Inc.
513 E. Thirty-sixth St.
Baltimore, MD 21218
(301) 243-4301

Dwyer Instruments Inc.
P.O. Box 373
Michigan City, IN 46360
(219) 872-9141

Dynamote Corp.
1200 W. Nickerson
Seattle, WA 98119
(206) 282-1000

Electro Sales Co., Inc.
100 Fellsway West
Somerville, MA 02145
(617) 666-0500

Elgar Corp.
8225 Mercury Ct.
San Diego, CA 92111
(619) 565-1155

Natural Power Inc.
Francestown Turnpike
New Boston, NH 03070
(603) 487-5512

Newark Electronics Corp.
500 N. Pulaski Rd.
Chicago, IL 60624
(312) 638-4411

Northwest Water Power Systems
P.O. Box 19183
Portland, OR 97219
(503) 288-1297

North Wind Power Co.
P.O. Box 556
Moretown, VT 05660
(802) 496-2955

Nova Electric Manufacturing Co.
263 Hillside Ave.
Nutley, NJ 07110
(201) 661-3434

PACS Industries, Inc.
P.O. Box 379
61 Steamboat Rd.
Great Neck, NY 11022
(516) 829-9060

Phoenix Power Systems, Inc.
1514 N.W. Forty-sixth St.
Seattle, WA 98107
(206) 784-1383

Satin American Corp.
P.O. Box 619
40 Oliver Terr.
Shelton, CT 06484
(203) 929-6363

Soleq Corp.
5969 N. Elston Ave.
Chicago, IL 60646
(312) 792-3811

Wilmore Electronics Co., Inc.
P.O. Box 1329
Hillsborough, NC 27278
(919) 732-9351

Windworks, Inc.
Box 44-A, Rt. 3
Mukwonago, WI 53149
(414) 363-4088

Wind-Driven Generators

Aeolian Energy, Inc.
R.D. 4
Ligonier, PA 15658
(412) 593-7905

Aerolite Inc.
P.O. Box 576
550 Russells Mills Rd.
South Dartmouth, MA 02748
(617) 993-9999

Aerowatt, S.A.
c/o Automatic Power, Inc.
P.O. Box 18738
Houston, TX 77223
(713) 228-5208

American Energy Savers, Inc.
912 St. Paul Rd.
Box 1421
Grand Island, NE 68802
(308) 382-1831

Astral/Wilcon Co. Inc.
(also known as A.W.I.)
P.O. Box 291
Millbury, MA 01527
(617) 865-9570

Bergey Wind Power Co., Inc.
2001 Priestly Ave.
Norman, OK 73069
(405) 364-4212

Bertoia Studio Ltd.
644 Main St.
Bally, PA 19503
(215) 845-7096

Bircher Machine, Inc.
Box 97
Kanopolis, KS 67454
(913) 472-4413

Carter Wind Systems, Inc.
Box 405-A, Rt. 1
Burkburnett, TX 76354
(817) 569-2238

DAF Indal Ltd.
3570 Hawkestone Rd.
Mississauga, ON L5C 2V8
Canada
(416) 272-5300

Dunlite Electrical Products
Enertech Corp.
P.O. Box 420
Norwich, VT 05055
(802) 649-1145

Elfin Corp.
550 Chippenhook Rd.
Wallingford, VT 05773
(802) 446-2575

Energy Science, Inc.
P.O. Box 3009
Boulder, CO 80307
(303) 449-3559

Enertech Corp.
P.O. Box 420
Norwich, VT 05055
(802) 649-1145

Environmental Energies, Inc.
Front St.
Copemish, MI 49625
(616) 378-2921
(616) 378-2922

Fayette Manufacturing
  Corporation
Box 1149
Tracy, CA 95376
(415) 443-2929

FloWind Corp.
21414 Sixty-eighth Ave., South
Kent, WA 98032
(206) 872-8500

Forecast Industries Inc.
3500 A Indian School Rd., NE
Albuquerque, NM 87106
(505) 265-3707

Jacobs Wind Electric
2720 Fernbrook Ln.
Minneapolis, MN 55401
(612) 559-9361

Kaman Aerospace Corp.
Old Windsor Rd.
Bloomfield, CT 06002
(203) 242-4461

Natural Energy Systems
  Unlimited
P.O. Box 60832
Rochester, NY 14606
(716) 458-9402

North Wind Power Co.
P.O. Box 556
Moretown, VT 05660
(802) 496-2955

Pinson Energy Corp.
Box 7
Marstons Mills, MA 02648
(617) 428-8535

PM Wind Power Inc.
P.O. Box 89
Mentor, OH 44060
(216) 255-3437

Power Group International, Inc.
12306 Rip Van Winkle
Houston, TX 77024
(713) 444-5000

Product Development Institute,
  Inc.
4445 Talmadge Rd.
Toledo, OH 43623
(419) 472-2136

Sencenbaugh Wind Electric
253 Polaris
Mountain View, CA 94043
(415) 964-1593

Tumac Industries, Inc.
650 Ford St.
Colorado Springs, CO 80915
(303) 596-4400

Winco Division of Dyna
  Technology
7850 Metro Pkwy.
Minneapolis, MN 55420
(612) 853-8400

Wind Engineering Corp.
P.O. Box 5936
Lubbock, TX 79417
(806) 763-3182

Windmaster Corp.
106 K St., Suite 200
Sacramento, CA 95814
(916) 443-0511

Wind Master Corp.
55 Veterans Blvd.
Carlstadt, NJ 07072
(201) 933-3338

Wind Power Systems, Inc.
8630 Production Ave.
San Diego, CA 92121
(619) 566-1806

Windstar Corp.
Box 151
Walled Lake, MI 48088
(313) 624-5597

Windworks, Inc.
Box 44A, Rt. 3
Mukwonago, WI 53149
(414) 363-4088

Winpower Corp.
1207 First Ave., East
Newton, IA 50208
(515) 792-1301

Zephyr Wind Dynamo Co., Inc.
Box 241
Brunswick, ME 04011
(207) 725-6534

Zond Systems, Inc.
Box 276
Tehachapi, CA 93561
(805) 822-6835

Publications

Cogeneration World (bimonthly)
International Cogeneration
   Society
Suite 1075
1111 Nineteenth St., NW
Washington, D.C. 20036
(202) 659-1552

The Cogenic Report (bimonthly)
Cogenic Energy Systems
645 Fifth Ave.
New York, NY 10022
(212) 832-6767

Hydrowire (biweekly)
Hydro Review (quarterly)
Hydro Consultants, Inc.
P.O. Box 344
Cambridge, MA 02138
(617) 491-5459

Photovoltaics (bimonthly)
Fore Publishers, Inc.
P.O. Box 3269
Scottsdale, AZ 85257
(602) 829-8167

Wind Power Digest (quarterly)
Wind Power Publishing, Inc.
P.O. Box 700
Bascom, OH 44809
(419) 937-2299

Appendix 7

States' Cogeneration Rate-Setting under PURPA

These listings represent the current status of PURPA-related rate setting as of May 1983. While the rates and terms listed below are specific for cogeneration systems, they are also relevant for photovoltaic, windpower and hydropower systems. For further information on all systems, you can contact your state Public Utilities Commission (PUC) or Public Service Commission (PSC).

STATE	STATUS	RATES
Ala.	Rates adopted for facilities producing 100 kw or less. Larger facilities negotiate with utilities.	**Ala. Power Co.:** For producers of 100 kw or less: Standard rate is 2.46¢/kwh through Oct., and 2.42¢/kwh Nov. through May. Time-of-day rate is 2.96¢/kwh peak and 2.46¢/kwh off-peak June through Oct., and 2.68¢/kwh peak and 2.42¢/kwh off-peak Nov. through May.
Alaska	Final rules issued. Utilities have begun filing proposed rates. Some rates have been approved; remainder should be approved this summer.	(for nonfirm producers of less than 100 kw) **Arctic Utilities** (approved): 11.40¢/kwh. **Kodiak Elec. Assn.** (approved): 7.591¢/kwh. **Chugach Elec. Assn. Inc.** (proposed): 0.685¢/kwh.
Ariz.	Final rules issued. Rates in effect but subject to Ariz. Corp. Com. investigation.	**Ariz. Pub. Ser. Co.:** Summer: 4.255¢/kwh peak, 1.986¢/kwh off-peak. Winter: 3.414¢/kwh peak, 2.028¢/kwh off-peak. Firm power suppliers receive an additional 10%.
Ark.	Final rules issued. Ark. Power & Light asked PSC to seek waiver of FERC rules requiring full avoided-cost payments, and has interim rates in effect not based on avoided costs.	**Ark. Power & Light Co.:** Summer: 3.531¢/kwh peak, 3.080¢/kwh off-peak. Winter: 3.127¢/kwh peak, 2.953¢/kwh off-peak.

NOTE: Abridged and reprinted from "States' Cogeneration Rate-Setting under PURPA, Part 1 and Part 2," *Energy User News*, 23 May 1983, pp. 16–17, and 30 May 1983, pp. 17–18, with permission of *Energy User News*, 7 E. Twelfth St., New York, NY 10003.

STATE	STATUS	RATES
Calif.	Final rules issued. Workshops to consider standard long-term contract offers are expected to be held in July.	Rates are reviewed quarterly. Capacity credits vary with length of contract; capacity credits listed are 20-yr contracts. **Pacific Gas & Elec.:** 4.40¢/kwh, capacity $110/kw/yr. **Southern Calif. Edison:** 4.02¢/kwh, capacity $114/kw/yr. **San Diego Gas & Elec.:** 6.45¢/kwh, capacity $93/kw/yr.
Colo.	Final rules issued. Utilities have filed proposed rates; PUC has begun proceeding to determine whether rates comply with rules.	**Pub. Ser. Co. of Colo.** (proposed): 1.77¢/kwh, capacity, $15.33/kw/mo.
Conn.	Final rules issued.	**Conn. Light & Power Co.:** Formula applies multipliers to utility's monthly average fossil fuel costs. Payments currently about 5¢/kwh. Firm power: 117% peak, 92% off-peak. Nonfirm power: 114% peak, 89% off-peak.
Del.	Final rules issued.	**Delmarva Power & Light Co.:** Summer: 6.37¢/kwh peak, 3.16¢/kwh off-peak. Winter: 6.25¢/kwh peak, 3.73¢/kwh off-peak.
D.C.	Final rules issued. Rates proposed.	**Potomac Elec. Power Co.:** Utility price offers range from 2.39 to 5.59¢/kwh.
Fla.	Rules issued, but new rules expected to be adopted by Aug., after PSC considers capacity payments. Recent Fla. court order overturning PSC rules has been appealed; rules remain in effect pending outcome of appeal.	For facilities producing over 100 kw: **Fla. Power Corp.:** 6.296¢/kwh peak, 4.665¢/kwh off-peak; **Fla. Power & Light Co.:** 5.029¢/kwh peak, 4.260¢/kwh off-peak; **Gulf Power Co.:** 3.523¢/kwh peak, 2.166¢/kwh off-peak; **Tampa Elec. Co.:** 4.201¢/kwh peak, 3.722¢/kwh off-peak. In addition, capacity payments may be negotiated if facility has annual availability of 70% or more.
Ga.	Final rules issued.	**Savannah Elec. Power Co.:** Summer: 4.082¢/kwh peak, 2.969¢/kwh off-peak. Winter: 2.683¢/kwh, all periods.
Hawaii	Final rules issued.	**Hawaiian Elec. Co.:** 6 to 6.60¢/kwh.
Idaho	Final rules and rates adopted. Idaho Power has asked PUC to lower payments and has filed lawsuit challenging PUC's authority to require and approve long-term contracts.	Power is sold under contract, with price updated annually to reflect coal prices. Capacity payments vary with length of contract; rates listed are for 20-yr contracts. **Idaho Power Co.:** 1.639¢/kwh; capacity, $232/kw/yr. **Wash. Water Power Co.:** 1.600¢/kwh; capacity, $202/kw/yr. **Utah Power & Light Co.:** 1.200¢/kwh; capacity, $188/kw/yr.

STATE	STATUS	RATES
Ill.	Final rules issued. Rates expected to be revised by 30 June.	**Commonwealth Edison:** Summer: 5.31¢/kwh peak, 2.90¢/kwh off-peak. Winter: 5.17¢/kwh peak. 3.37¢/kwh off-peak. **Iowa-Ill. Gas & Elec.:** Summer: 2.30¢/kwh peak, 1.19¢/kwh off-peak. Winter: 2.38¢/kwh peak, 1.28¢/kwh off-peak. **South Beloit Water, Gas & Elec. Co.:** 2.65¢/kwh peak, 1.93¢/kwh off-peak.
Ind.	Rules and rates approved, but may change after hearings in May.	**Northern Ind. Pub. Ser. Co.** Rates range from 2.46¢/kwh winter off-peak to 3.33¢/kwh summer peak. **Pub. Ser. Co. of Ind.:** 1.330¢/kwh.
Iowa	Final rules issued; rates now being investigated.	**Iowa Pub. Ser. Co.:** Option A: 1.82¢/kwh. Option B: 2.03¢/kwh peak, 1.64¢/kwh off-peak. **Iowa-Ill. Gas & Elec:** Option A: 1.88¢/kwh summer, 1.63¢/kwh winter. Option B: Rates range from 1.29¢/kwh winter off-peak to 2.55¢/kwh summer peak. Capacity payment determined by formula involving kwh delivered during peak, and utility's cost to borrow power from power pool.
Kans.	Kans. Corp. Com. issued final rules requiring utilities to file rates. Kans. City Power & Light Co. lawsuit challenging the authority of the Kans. Corp. Com. to require long-term contracts expected to be heard by State Supreme Court.	**Kans. Gas & Elec. Co.:** Rates equal to utility's total cost of fuel and purchased power as determined by monthly energy adjustment clause. Average rates currently 2 to 2.5¢/kwh. Capacity payments determined by formula.
Ky.	Final rules issued. Rates proposed.	**Ky. Utilities Co.:** 1.5¢/kwh. **Louisville Gas & Elec.:** 1.7¢/kwh.
La.	Final rules issued. Utilities must file standard tariff rates and avoided costs by late May.	Not available.
Maine	Final rules issued.	Revised periodically in fuel adjustment proceedings. Rates listed are for June. **Central Maine Power Co.:** 4.6¢/kwh peak, 3.7¢/kwh off-peak. **Bangor Hydroelec. Co.** (for producers of 1,000 kw or less): 4.0¢/kwh peak, 3.3¢/kwh off-peak.
Md.	Final rules issued. Filings approved for most utilities; further hearings to be held	**Potomac Elec. Power Co.:** Rates range from 2.641¢/kwh for winter off-peak to 5.276¢/kwh for summer peak. **Potomac**

STATE	STATUS	RATES
Md. (continued)	to consider capacity payments.	**Md. Edison Co.:** 1.57¢/kwh. **Delmarva Power & Light Co.:** Rates range from 2.52¢/kwh winter off-peak to 4.22¢/kwh summer peak. **Baltimore Gas & Elec. Co.:** Rates range from 1.98¢/kwh winter off-peak to 5.97¢/kwh summer peak.
Mass.	Final rules issued. Rates proposed.	**Western Mass. Elec.:** 5.06¢/kwh peak and 4.36¢/kwh off-peak, or a flat 4.70¢/kwh. **Mass. Elec.:** 4.92¢/kwh peak and 3.32¢/kwh off-peak, or a flat 4.07¢/kwh. **Boston Edison Co.:** 5.22¢/kwh peak and 3.79¢/kwh off-peak, or a flat 4.45¢/kwh.
Mich.	Final rules issued.	For nonfirm power: 3.0¢/kwh. For firm power: 5 to 6¢/kwh (including capacity payment).
Minn.	Final rules issued. Utilities have until 24 June to file rates.	**Northern States Power** (100 kw or less): Option A: 2.67¢/kwh peak, 1.66¢/kwh off-peak. Under Option B, rates depend on length of contract signed, and are adjusted each yr. Option B rates for 1982 range from 2.28¢/kwh for a 5-yr contract to 3.40¢/kwh for a 25-yr contract.
Miss.	No rules issued or rates approved.	Not available.
Mo.	Final rules issued. Utilities filed rates ranging from 1.5 to 5.0¢/kwh, which have been suspended pending an investigation. Only Kans. City Power & Light Co.'s rates have been approved to date.	**Kans. City Power & Light Co.** (for nonfirm producers of under 100 kw): 1.63¢/kwh.
Mont.	PSC issued final rules, but will review rules and rate methodology in hearings beginning 21 June. Mont. Power Co. filed motion to dismiss new proceeding, saying PSC lacks authority.	(Long-term rate requires 4-yr contract.) **Mont. Power Co.:** 2.34¢/kwh short-term power, 4.09¢/kwh long-term power, 6.74¢/kwh peak. **Pacific Power & Light Co.:** 7.76¢/kwh short-term peak, 1.84¢/kwh short-term off-peak, 6.05¢/kwh long-term peak, 4.03¢/kwh long-term off-peak. **Mont.-Dak. Utilities Co.:** 2.16¢/kwh short-term, 5.58¢/kwh long-term, 5.53¢/kwh peak.
Nebr.	Nebr. has 166 municipal and 4 cooperative utilities, which are not regulated by any state body and set their own cogeneration rates and rules.	**Omaha Pub. Dist.** (for producers of 100 kw or less): Summer: 1.60¢/kwh peak and 1.00¢/kwh off-peak, or flat 1.10¢/kwh. Winter: 1.20¢/kwh peak and 1.00¢/kwh off-peak, or flat 1.10¢/kwh.

STATE	STATUS	RATES
Nev.	Rules issued.	**Nev. Power Co.:** Summer: 3.81¢/kwh peak, 2.33¢/kwh off-peak. Winter: 2.99¢/kwh peak, 2.27¢/kwh off-peak. Capacity payments available to reliable facilities that meet certain conditions. Capacity payment $7.65/kw peak, 24¢/kw off-peak.
N.H.	Rules and rates issued, but rate revision hearings to be held this summer.	7.7¢/kwh. 8.2¢/kwh for reliable facilities (7.7¢/kwh for energy plus 0.5¢/kwh for capacity).
N.J.	Final rules issued. Further hearings may be held.	Energy payments equal 110% of rate utilities pay for power from Pa.-N.J.-Md. power pool. Capacity payments also tied to power pool rates. **Pub. Ser. Elec. & Gas Co.:** rates range from 3.697¢/kwh winter off-peak to 8.14¢/kwh summer peak. Capacity, $30.66/kw/yr. **Atlantic City Elec. Co.:** 5.36¢/kwh guaranteed minimum. Capacity, $30.66/kw/yr.
N.Mex.	Rules issued.	(for producers of 100 kw or less) **Pub. Ser. Co. of N. Mex.:** Primary voltage (power bought at 12.6 kv): Summer: 4.89¢/kwh peak and 2.99¢/kwh off-peak, or flat 3.47¢/kwh. Winter: 4.95¢/kwh peak and 4.05¢/kwh off-peak or flat 3.83¢/kwh. Secondary voltage (power bought at less than 12.6 kv): Summer: 5.23¢/kwh peak and 3.23¢/kwh off-peak, or flat 3.66¢/kwh. Winter: 5.24¢/kwh peak and 4.33¢/kwh off-peak, or flat 4.01¢/kwh.
N.Y.	PSC issued final order in case of Con Ed Co. of N.Y. Inc. and approved Con Ed's rates. Other utilities filed rates based on this order; other utilities' rates expected to be approved by end of summer. Con Ed is challenging PSC rules in state court.	State law sets minimum average of 6¢/kwh. Rates listed are proposed, except for Con Ed's, which are approved. **Con Ed** (over 900 kw, low-tension service): Summer: 12.37¢/kwh peak, 4.77¢/kwh off-peak. Winter: 6.57¢/kwh peak, 4.37¢/kwh off-peak. **Niagara Mohawk Power:** 5.46¢/kwh peak (includes capacity payment), 4.00¢/kwh off-peak. **Orange & Rockland Utilities:** Summer: 6.733¢/kwh peak (includes capacity payment), 4.179¢/kwh off-peak. Winter: 5.408¢/kwh peak, 4.836¢/kwh off-peak. **Central Hudson Gas & Elec.** (rates include capacity payments): Secondary voltage: 9.98¢/kwh summer peak, 7.57¢/kwh winter peak, 5.84¢/kwh spring/fall peak, 4.29¢/kwh off-peak. Primary voltage: 10.64¢/kwh summer peak, 8.10¢/kwh winter peak, 6.27¢/kwh spring/fall peak, 4.58¢/kwh off-peak. Transmission

STATE	STATUS	RATES
N.Y. (continued)		voltage: 10.12¢/kwh summer peak, 7.68¢/kwh winter peak, 5.92¢/kwh spring/fall peak, 4.45¢/kwh off-peak.
N.C.	Final rules issued. Proposed new rates now being reviewed; approval expected in June.	(proposed) **Carolina Power & Light Co.:** 4.25¢/kwh peak, 2.73¢/kwh off-peak. 5-yr contract: 4.93¢/kwh peak, 3.08¢/kwh off-peak. 10-yr contract: 5.98¢/kwh peak, 3.62¢/kwh off-peak. 15-yr contract: 7.68¢/kwh peak, 4.49¢/kwh off-peak. Capacity payments based on kwh supplied during peak hours. Capacity: 5-yr or 10-yr contract: 2.15¢/kwh summer peak, 1.86¢/kwh nonsummer peak. 15-yr contract: 3.63¢/kwh summer peak, 3.14¢/kwh nonsummer peak. **Duke Power Co.:** 3.03¢/kwh peak, 2.13¢/kwh off-peak. 5-yr contract: 3.27¢/kwh peak, 2.25¢/kwh off-peak. 10-yr contract: 4.08¢/kwh peak, 2.63¢/kwh off-peak. 15-yr contract: 4.79¢/kwh peak, 2.95¢/kwh off-peak. Capacity payment: 1.27¢/kwh summer peak, 0.77¢/kwh nonsummer peak. 5-yr or 10-yr contract: 1.39¢/kwh summer peak, 0.84¢/kwh nonsummer peak. Contracts for 11 yr or longer: 1.83¢/kwh summer peak, 1.10¢/kwh nonsummer peak.
N.Dak.	Final rules issued.	For cogenerators producing over 500 kw: 1.6 to 3.0¢/kwh.
Ohio	Interim rules issued. Utilities have proposed rates, which will be investigated by PUC.	**Ohio Power Co.:** 1.70 to 2.00¢/kwh.
Okla.	Final rules issued. Interim rates approved for facilities producing under 100 kw; proceedings initiated to determine appropriate avoided-cost methodology; hearings expected.	(interim rates for 100% reliable facilities) **Pub. Ser. Co. of Okla.:** 4.054¢/kwh, 2.744¢/kwh off-peak. **Okla. Gas & Elec.:** 3.308¢/kwh.
Oreg.	Final rules issued. Utilities filed rates, but are required by 1 July to file rates based on avoided costs.	3.8 to 4.5¢/kwh.
Pa.	Final rules issued. Utilities have proposed rates. Four utilities have challenged PUC rules in state court.	**Pa. Power & Light Co.:** 6¢/kwh to cogenerators who use only renewable resources.

STATE	STATUS	RATES
R.I.	Final order issued.	**Narragansett Elec. Co.:** 5.38¢/kwh peak and 4.038¢/kwh off-peak, or flat 4.736¢/kwh. **Blackstone Valley Elec. Co.:** Primary voltage: 5.643¢/kwh peak and 4.293¢/kwh off-peak, or flat 4.871¢/kwh. Secondary voltage: 5.920¢/kwh peak and 4.420¢/kwh off-peak, or flat 5.058¢/kwh. **Newport Elec. Co.:** 4.54¢/kwh peak and 4.14¢/kwh off-peak, or flat 4.38¢/kwh. **Pascoag Fire Dist:** 3.085¢/kwh peak, 2.902¢/kwh off-peak. **Block Island Power Co.:** 14.465¢/kwh.
S.C.	Final rules issued. Rates approved.	**S.C. Elec. & Gas Co.:** 3.26¢/kwh peak, 2.275¢/kwh off-peak. Capacity, $2.75/kw/mo. **Carolina Power & Light Co.:** 2.80¢/kwh peak, 2.07¢/kwh off-peak. Capacity, $3.89/kw/mo July through Oct., $3.35/kw/mo Nov. through June. **Duke Power Co.:** 1.9¢/kwh peak, 1.49¢/kwh off-peak. Capacity, $5.00/kwmo.
S.Dak.	Final rules issued. Rates approved for facilities producing 100 kw or less.	**Black Hills Power & Light Co.:** 1.30 to 3.50¢/kwh.
Tenn.	Hearing held to consider rates filed by utility. Final order not expected soon.	(proposed) **Kingsport Power Co.:** 1.36¢/kwh peak and 0.81¢/kwh off-peak or flat 0.81¢/kwh. Capacity, $3.00/kw/mo if on time-of-day metering, otherwise $1.50/kw/mo.
TVA	Experimental rates and interim rules extended to Oct. New rules issued for purchase of power from cogenerators outside TVA area.	(interim) 4.64¢/kwh peak and 2.92¢/kwh off-peak, or flat 3.44¢/kwh.
Tex.	Final rules issued. Utilities have filed rates. Houston Lighting & Power Co. has asked for decrease in rates.	**Houston Lighting & Power Co.:** Option A: Multipliers applied to utility's average fuel cost, currently about 3.0¢/kwh. For non-PURPA QFs, multipliers are 1.31 summer peak, 1.01 summer off-peak, 1.13 winter. For QFs, multipliers are 1.64 summer peak, 1.27 summer off-peak, 1.42 winter. Option B (avail. to QFs over 5,000 kw): hourly payment based on avoided cost formula.
Utah	Final rules issued. Hearing to review policy and revise rates scheduled for 18 July.	**Utah Power & Light Co.:** (for producers of 1,000 kw or less): 2.2¢/kwh nonfirm, 2.6¢/kwh firm.

STATE	STATUS	RATES
Vt.	Final rules issued. Hearings on revised rate expected to be held in July.	9¢/kwh peak and 6.6¢/kwh off-peak, or flat 7.8¢/kwh.
Va.	Final rules issued. Rates approved.	**Va. Elec. & Power Co.:** 5.203¢/kwh peak, 3.132¢/kwh off-peak. Capacity: 0.803¢/kwh for facilities operating less than 5 yr. **Appalachian Power Co.:** Rates range from 1.36¢/kwh winter off-peak to 1.7¢/kwh summer peak. Capacity: $3.00/kw/mo seasonal peak, $1.50/kw/mo off-peak.
Wash.	Final rules issued. Rates approved.	**Puget Sound Power & Light Co.:** 2.07¢/kwh. **Pacific Power & Light:** 2.795¢/kwh.
W.Va.	Rules proposed. Issuance of final rules had been awaiting outcome of Am. Elec. Power lawsuit.	(proposed) **Monogahela Power Co.:** 1.00 to 2.00¢/kwh.
Wis.	Rates in effect. Hearings continuing. Final rules expected in June.	**Wis. Power & Light Co.:** 4.8¢/kwh peak, 1.75¢/kwh off-peak. **Madison Gas & Elec. Co.:** 2.75¢/kwh summer peak, 2.22¢/kwh winter peak, 1.50¢/kwh off-peak. **Wis. Elec. Power:** 3.65¢/kwh summer peak, 3.45¢/kwh winter peak, 1.45¢/kwh off-peak. **Northern States Power Co.:** 1.60¢/kwh peak, 1.14¢/kwh off-peak. Capacity: $4.00/kw/mo. **Lake Superior Dist. Power Co.:** 1.90¢/kwh. Capacity: $6.02/kw/mo. **Wis. Pub. Ser. Corp.:** 1.85¢/kwh peak, 1.32¢/kwh off-peak. Capacity payments determined by each facility's degree of firmness.
Wyo.	Final rules issued.	**Cheyenne Light, Fuel & Power Co.** (for nonfirm producers of 100 kw or less): 4.05¢/kwh.

NOTES

Chapter 1 The Electric Revolution
1. John A. Kuecken, *How to Make Electricity from Wind, Water and Sunshine,* p. 8.
2. Sheldon Novick, "The Electric Power Industry," *Environment,* November 1975, p. 34.
3. Margaret Cheney, *Tesla,* p. 24.
4. Novick, "Electric Power Industry," p. 35.
5. Stiles P. Jones, "State Versus Local Regulation," *The Annals,* May 1914, p. 103.
6. Martin G. Glaeser, *Public Utilities in American Capitalism* (New York: Macmillan Publishing, 1957), pp. 120–21.
7. *Congressional Record,* 66th Congress, 1925, p. 1974.
8. Glaeser, *Public Utilities,* p. 152.
9. Jeanne Hardy, "Electricity Comes to Washington," *Solar Washington,* July/August 1982.
10. William F. Cafron, ed., *Technological Change in Regulated Industries* (Washington, D.C.: Brookings Institution, 1971), p. 54.
11. Amory B. Lovins and L. Hunter Lovins, "Energy Policies for Resilience and National Security," unpublished draft, 1981, p. 49.
12. Ibid., p. 50.
13. Amory B. Lovins and L. Hunter Lovins, *Brittle Power,* p. 128.
14. J. W. Hilborn and J. S. Glen, "Small Nuclear Reactors," *Energy Forum,* Spring 1982.
15. Office of Technology Assessment, *Application of Solar Technology to Today's Energy Needs,* vol. I (Washington, D.C.: Office of Technology Assessment, 1978).
16. U.S. Congress, House, Committee on Interstate and Foreign Commerce, *Local Energy Policies Hearings before the Subcommittee on Energy and Power,* 95th Congress, 5–9 June 1978.
17. John McPhee, "Minihydro," *The New Yorker,* 23 February 1981, p. 45.
18. Idaho Public Utilities Commission, "In the Matter of Rule Making Proceedings as Required by the Public Utility Regulatory Policies Act of 1978 for the Consideration of Cogeneration and Small Power Production." case no. P–300–12, order no. 15746, 8 August 1980.
19. U.S. District Court, Southern District of Mississippi, Jackson Division, Civil Action J–79–0212, State of Mississippi et al. v. FERC et al., Mississippi Power and Light Company.

Chapter 2 How the Electric System Works
1. Philip Boffey, "Investigators Agree N.Y. Blackout of 1977 Could Have Been Avoided," *Science*, vol. 201, 15 September 1978, pp. 994–98. Copyright 1978 by the American Association for the Advancement of Science. Reprinted by permission of the American Association for the Advancement of Science.
2. Charles Komanoff, *Power Plant Cost Escalation* (New York: Komanoff Energy Associates, 1981).
3. Amory B. Lovins and L. Hunter Lovins, *Brittle Power*, p. 341.
4. Ibid.
5. Edward Kahn, private communication with the author, 1982.
6. Andrew Ford and Irving Yabroff, "Defending against Uncertainty in the Electric Utility Industry," *LA-UR-783228*, December 1978.
7. Edward Kahn, "Project Lead Times and Demand Uncertainty: Implications for the Financial Risk of Utilities," paper presented at E. F. Hutton Fixed Income Research Conference on Electric Utilities, 8 March 1979, pp. 2–4.
8. "Utilities: Weak Point in the Energy Future," *Business Week*, 20 January 1975, p. 46.

Chapter 3 PURPA
1. Peter Brown testifying before U.S. Congress, House, Subcommittee on Energy Conservation and Power of the Energy and Commerce Committee, Hearings on HR6500, 97th Congress, 15 June 1982.
2. "Small Power Production and Cogeneration Facilities; Regulations Implementing Section 210 of the Public Utility Regulatory Policies Act of 1978," *Federal Register* 45, no. 38 (25 February 1980), p. 12226.
3. Before the Public Service Commission of Utah, case no. 80-999-06, 9 April 1981.
4. "Making Interconnection Work: A Special Report," *Power*, June 1982, pp. 8–14.
5. Thaxter Lipex Stevens Broder & Micoleau, *Key Provisions of a Sample Purchase Power Agreement*, p. 3.
6. *Public Utility Regulatory Policies Act of 1978*, Title II, Section 210(d).
7. "Small Power Production," *Federal Register*, p. 12222.
8. Public Utilities Commission of California, decision no. 82-01-103, 21 January 1982.
9. "Small Power Production," *Federal Register*, p. 12216.

10. Before the Public Utilities Commission of New Hampshire, case no. DE-79-208, 18 June 1980.
11. "Small Power Production," *Federal Register*, p. 12224.
12. Ibid.
13. Ibid., p. 12225.
14. California Public Utilities Commission Final Staff Report on Cogeneration and Small Power Production Pricing Standards, OIR 2 (San Francisco: California Public Utilities Commission, 20 January 1981), p. 8.7.
15. Recommended decision of Hearing Examiner Michael R. Homyak, case no. 5970, Public Utilities Commission of Colorado, decision no. R81-801, 6 May 1981.
16. Compliance Hearing on Standard Offers and Tariffs Filed in Response to Orders in OIR-2 (San Francisco: California Public Utilities Commission, 30 September 1982).
17. Paul A. London and Alan L. Madian, Analysis and Recommendations Regarding Public Service Commission's Order 82-10: Niagara Mohawk Submission of Draft P.S.C. no. 207 Electricity, Service Classification no. 11., before the State of New York Public Service Commission, case no. 27574, August 1982.
18. Ibid.
19. Ibid.
20. New Jersey Board of Public Utilities, formal case no. 757, decision and order docket no. 8010-687, 14 October 1981.
21. Testimony of Nixon and Trotter before the Arkansas Public Service Commission, 1982.
22. Before the Public Utility Commissioner of Oregon, "In the Matter of Establishing Base Rates for Cogeneration and Small Power Production under ORS 758.530(2)," order 82–515, 20 July 1982, p. 3.
23. "Small Power Production," *Federal Register*, p. 12220.
24. Otter Tail Power Company v. U.S. 410 U.S. 359 (1973).
25. J. Richard Tiano and Michael J. Zimmer, "Wheeling for Cogeneration and Small Power Facilities," *Energy Law Journal* 3 (1981), p. 97.
26. Florida Power and Light Company v. FERC 660 F. 2d 668, 5th Circuit (1981).
27. Federal Power Commission v. Southern California Edison Company, 376 U.S. 205, denied 277 U.S. 913 (1964).
28. California Public Utilities Commission Decision 82-01-103 (3 September 1980), p. 110.
29. Minnesota H.F. no. 473.

30. Testimony of James Donald Hebson, Jr., before New Jersey Board of Public Utilities, docket no. 8010-687 phase II.
31. Testimony of Edward P. Kahn before New Jersey Board of Public Utilities, docket no. 8010-687 phase II.

Chapter 4 Interconnecting with the Grid

1. This chapter draws heavily from Howard S. Geller, *The Interconnection of Cogenerators and Small Power Producers to a Utility System.*
2. David Curtice and James Patton, *Operation of Small Wind Turbines on a Distribution System.*
3. "Making Interconnection Work: A Special Report," *Power*, June 1982, pp. 5–6.
4. Ibid., pp. 5–16.
5. Ibid.

Chapter 5 Getting the Best Deal

1. Thaxter Lipex Stevens Broder & Micoleau, *Key Provisions of a Sample Purchase Power Agreement*, p. 3.
2. Ibid.
3. Before the Public Service Commission of the State of Montana, docket no. 81.2.15, order no. 4865a, 18 February 1982.
4. Ibid.
5. Ibid.
6. See application of Pacific Gas and Electric Company for an order approving certain provisions of a power sales agreement between U.S. Windpower, Inc., and Pacific Gas and Electric Company, before the Public Utilities Commission of California, 23 November 1981.
7. Thaxter, *Key Provisions*, p. 22.
8. Draft copy of agreement between New York State Electric and Gas Corporation and unidentified wind turbine owner, 1982.
9. Ibid.
10. Thaxter, *Key Provisions*, p. 24.
11. "Small Power Production and Cogeneration Facilities; Regulations Implementing, Section 210 of the Public Utility Regulatory Policies Act of 1978," *Federal Register* 45, no. 38 (25 February 1980), p. 12230.

Chapter 6 Electric Generation Technologies

1. E. F. Lindsley, "33 windmills you can buy now!" *Popular Science*, July 1982, p. 54.

2. Ibid.
3. Paul Gipe, *Fundamentals of Wind Energy*, p. 23.
4. Lindsley, "33 windmills," p. 54.
5. Jim Leckie, Gil Masters, Harry Whitehouse and Lily Young, *More Other Homes and Garbage*, p. 45.
6. Ron Alward, Sherry Eisenbart and John Volkman, *Micro-Hydro Power*, p. 21.
7. Daniel J. Schneider and Emory K. Damstrom, *The Schneider Engine*, p. 3.
8. J. George Butler, *How to Build and Operate Your Own Small Hydroelectric Plant*, p. 88.

Chapter 7 Economics
1. Ned Coffin, "The 'My Own Electric Company': How to Make Money While Getting Even With Uncle Sam and Your Local Utility Company," *Alternative Sources of Energy* (59): pp. 21–22.

BIBLIOGRAPHY

Alternative Energy Section. *Handbook of Cogeneration Pricing Methodologies.* San Francisco: California Public Utilities Commission, 1981.

Alward, Ron; Eisenbart, Sherry; and Volkman, John. *Micro-Hydro Power: Reviewing an Old Concept.* Butte, Mont.: The National Center for Appropriate Technology, 1979.

American Gas Association. *A.G.A. Marketing Manual Natural Gas Prime Movers 1981.* Washington, D.C.: American Gas Association, 1981.

Brown, Peter W. *The Financing of Private Small Scale Hydroelectric Projects.* Concord, N.H.: Energy Law Institute, Franklin Pierce Law Center, 1981.

————. *Legal Obstacles and Incentives to Small Scale Hydroelectric Development in the Six Middle Atlantic States.* Concord, N.H.: Energy Law Institute, Franklin Pierce Law Center, 1979.

Brown, Robert J., and Yanuck, Rudolph R. *Life Cycle Costing: A Practical Guide for Energy Managers.* Atlanta: The Fairmont Press, 1980.

Brown, Ruben S., and Goodman, Alvin S. *Site Owner's Manual for Small Scale Hydropower Development in New York State, Report 79-3.* Albany: New York State Energy Research and Development Authority, 1980.

Bullard, Clark W., and Pien, Shyh-Jye. *Residential and Commercial Cogeneration Systems Assessment.* Chicago: Gas Research Institute, 1982.

Butler, J. George. *How to Build and Operate Your Own Small Hydroelectric Plant.* Blue Ridge Summit, Pa.: TAB Books, 1982.

Caskey, David L.; Caskey, Bill C.; and Aronson, Eugene A. *Parametric Analysis of Residential Grid-Connected Photovoltaic Systems with Storage.* Albuquerque: Sandia National Laboratories, 1980. Order no. SAND79-2331.

Caterpillar Engine Division. *On Site Power Generation*, Caterpillar Engine Division, n.d.

Cheney, Margaret. *Tesla: Man Out of Time*. Englewood Cliffs, N.J.: Prentice-Hall, 1981.

Colorado Office of Energy Conservation. *Water Over the Dam: A Small Scale Hydro Workbook for Colorado*. Denver: Colorado Office of Energy Conservation, July 1981, revised January 1982.

Copperfield, David. *How To Install Solar Electric Panels*. Camptonville, Calif.: Electrical Independence Booklets, 1982.

Cost Estimating Guidebook for Interconnections Between Electric Utilities and Small Power Producers Qualifying Under PURPA. Draft. Washington, D.C.: U.S. Department of Energy, 1982.

"Creating the Electric Age: Roots of Industrial R & D." *EPRI Journal*, March 1979, entire issue.

Cullen, Jim. *How To Be Your Own Power Company: The Low-Voltage, Direct-Current, Power-Generating System*. New York: Van Nostrand Reinhold, 1980.

Curtice, David and Patton, James. *Operation of Small Wind Turbines on a Distribution System*. Palo Alto: Systems Control, 1981. Order no. RFP-3177-2.

Danzinger, Robert N.; Caples, Patrick W.; and Huning, James R. *The Rules Implementing Sections 201 and 210 of the Public Utility Regulatory Policies Act of 1978: A Regulatory History*. Pasadena: Jet Propulsion Laboratory, 1980. Order no. DOE/JPL-1012-48.

Davitian, Harry, and Brainard, Joel. *Purchases from Qualifying Facilities under PURPA 210: Guidelines for Developing Rates and Designing Tariffs*. Golden, Colo.: Colorado Energy Research Institute, 1982.

Deshmukh, R. G., and Ramakumar, R. *Reliability of Wind-Assisted Utility Systems*. Stillwater, Okla.: Oklahoma State University, 1979.

Elliott, D. L.; Barchet, W. R.; and George, R. L. *Wind Energy Resource Atlas*. 12 vols. Richland, Wash.: Battelle Pacific Northwest

Laboratory, 1980. Available from National Technical Information Service, Springfield, VA 22161. Order no. PNL 3195 WERA-4.

Elridge, Frank. *Wind Machines.* New York: Van Nostrand Reinhold, 1980.

Energy Primer: Solar, Water, Wind and Biofuels. Menlo Park, Calif.: Portola Institute, 1974.

Enertech Corporation. *Planning a Wind-Powered Generating System.* Norwich, Vt.: Enertech Corporation, 1977.

Feuerstein, Randall J. *Utility Rates and Service Policies as Potential Barriers to the Market Penetration of Decentralized Solar Technologies.* Golden, Colo.: Solar Energy Research Institute, 1979.

Feurer, Duane A., and Weaver, Clifford L. *Study of the Impacts of Regulations Affecting the Acceptance of Integrated Community Energy Systems: Background Report United States.* Washington, D.C.: U.S. Department of Energy. Order no. EM-78-C-02-4627.

Finder, Alan E. *The States and Electric Utility Regulation.* Lexington, Ky.: The Council of State Governments, 1977.

Flaim, Theresa; Considine, Timothy J.; Witholder, Robert; and Edesess, Michael. *Economic Assessments of Intermittent, Grid-Connected Solar Electric Technologies: A Review of Methods.* Golden, Colo.: Solar Energy Research Institute, 1981.

Foley, Michael. *Discussion Report on the Current Financial Condition of the Investor Owned Electric Utility Industry.* Washington, D.C.: National Association of Regulatory Utility Commissioners, 1982.

Geller, Howard S. *The Interconnection of Cogenerators and Small Power Producers to a Utility System: Equipment Costs.* Report to the Office of the People's Counsel. Washington, D.C.: Self-Reliance, Inc., 1982.

Gipe, Paul. *Adopt an Anemometer.* Harrisburg, Pa.: Center for Alternate and Renewable Energy, 1982.

————. *Fundamentals of Wind Energy.* Harrisburg: Paul Gipe, 1981. Available from the author, P.O. Box 539, Harrisburg, PA 17108.

————. *Wind Potential in Pennsylvania*. Harrisburg: Pennsylvania Department of Community Affairs, 1979.

Glaser, Martin G. *Public Utilities in American Capitalism*. New York: Macmillan Publishing, 1957.

Gordon, Richard L. *Reforming the Regulation of Electric Utilities*. Lexington, Mass.: Lexington Books, 1982.

Governor's Commission on Cogeneration. *Cogeneration: Its Benefits to New England*. Boston: Massachusetts Energy Office, 1978.

Gupta, Y.; Knowles, R.; Merrill, O.; and Young, S. "Economic Worth of On-Site Solar Electric Generation in a Utility Grid." In *Proceedings of the Second Annual Systems Simulation and Economic Analysis Conference*, Golden, Colo.: Solar Energy Research Institute, 1980. Order no. SERI/TP-351-431.

Hayes, Gail Boyer. *Solar Access Law: Protecting Access to Sunlight for Solar Energy Systems*. Cambridge: Ballinger Publishing, 1979.

Huetter, John Jr. *Design of Low Cost, Ultra-Low Head Hydropower Package Based on Marine Thrusters*. Santa Monica: Energy Research and Applications, 1981.

Hydrologic Engineering Center. *Feasibility Studies for Small Scale Hydropower Additions: A Guide Manual*. Springfield, Va.: National Technical Information Service, 1979. Order no. DOERA0048.

Intertechnology/Solar Corporation. *Photovoltaic Power Systems Market Identification and Analysis*. Washington, D.C.: U.S. Department of Energy, 1979.

James, Jeffrey W., and McCoy, Gilbert A. *Developing Hydropower in Washington State: An Electricity Marketing Manual Volumes I and II*. Olympia: Washington State Energy Office, 1982.

Johanson, E. E. "Synthesis of WECS/Utility Integration Studies: Centralized and Dispersed," In *Proceedings of the Fourth Biennial Conference and Workshop on Wind Energy Conversion Systems*. Washington, D.C.: U.S. Department of Energy, 1980. Order no. CONF-791097.

Juhasz, Paul. *Cogeneration Handbook*. Olympia: Washington State Energy Office, 1982.

Justus, C. G. *Winds and Wind System Performance*. Philadelphia: Franklin Institute, 1978.

Komp, Richard. *Practical Photovoltaics*. English, Ind.: Skyheat Associates, 1982.

Kuecken, John A. *How to Make Home Electricity from Wind, Water & Sunshine*. Blue Ridge Summit, Pa.: TAB Books, 1980.

Lacy, Edward A. *How to Cut Your Electric Bill and Install Your Own Emergency Power System*. Blue Ridge Summit, Pa.: TAB Books, 1978.

Leckie, Jim; Masters, Gil; Whitehouse, Harry; and Young, Lily. *More Other Homes and Garbage*. San Francisco: Sierra Club Books, 1981.

Loehr, William. *Guide to Negotiations between Small Power Producers and Utilities*. Denver, Colo.: Colorado Small Hydro Office, 1982.

Lovins, Amory B., and Lovins, L. Hunter. *Brittle Power*. Andover, Mass.: Brick House Publishing Company, 1982.

McGuigan, Dermot. *Harnessing the Wind for Home Energy*. Charlotte, Vt.: Garden Way Publishing Company, 1978.

————. *Harnessing Water Power for Home Energy*. Charlotte, Vt.: Garden Way Publishing Company, 1978.

MacKay, Robin. *Gas Turbines and Cogeneration*. Los Angeles: The Garrett Corporation, 1979.

McLaughlin, Terence. *Make Your Own Electricity*. North Pomfret, Vt.: David and Charles, 1977.

Marier, Donald. *Wind Power for the Homeowner*. Emmaus, Pa.: Rodale Press, 1981.

Marsh, W. D. *Requirements Assessment of Photovoltaic Power Plants in Electric Utility Systems*. Palo Alto: Electric Power Research Institute, 1978.

Maycock, Paul D., and Stirewalt, Edward N. *Photovoltaics: Sunlight to Electricity in One Step*. Andover, Mass.: Brick House Publishing Company, 1981.

Mehalick, E. M.; O'Brien, G.; Tully, G. F.; Johnson, J.; and Parker, J. *The Design of a Photovoltaic System for a Southwest All-Electric Residence*. Philadelphia: General Electric Space Division, 1980.

Merrigan, Joseph A. *Sunlight to Electricity: Photovoltaic Technology and Business Prospects*. Cambridge, Mass.: MIT Press, 1982.

Messing, Marc; Friesema, H. Paul; and Morell, David. *Centralized Power: The Politics of Scale in Electricity Generation*. Cambridge: Oelgeschlager, Gunn & Hain, Publishers, 1979.

Monegon, Ltd. *Solar Electricity: Making the Sun Work for You*. Gaithersburg, Md.: Monegon, 1981.

Morris, David. *Self-Reliant Cities: Energy and the Transformation of Urban America*. San Francisco: Sierra Club Books, 1982.

Office of Technology Assessment. *Application of Solar Technology to Today's Energy Needs*. Vols. 1 and 2. Washington, D.C.: U.S. Government Printing Office, 1978. Order no. 052-003-00608-1.

Oregon Department of Energy. *Oregon Wind Book*. Salem: Oregon Department of Energy, 1980.

Osborn, William C., and Downey, William T. *Near-Term High Potential Counties for SWECS*. Golden, Colo.: Solar Energy Research Institute, 1981. Order no. SERI/TR-98282-11.

Park, Jack. *The Wind Power Book*. Palo Alto: Cheshire Books, 1981.

Park, Jack, and Obermeier, John. *Common Sense Wind Energy*. Sacramento: Office of Appropriate Technology, 1981.

Park, Jack, and Schwind, Richard. *Wind Power for Farms, Homes and Small Industry*. Springfield, Va.: National Technical Information System, 1977. Order no. RFP 2841/1270/78/4.

Patton, J. B. *Survey of Utility Cogeneration Interconnection Practices and Cost.* Washington, D.C.: U.S. Department of Energy, 1980. Order no. DOE/RA/29349-01.

Patton, J. B. and Iqbal, S. *Small Power Production Interconnection Issues and Costs.* Palo Alto: Systems Control, 1981.

Paul, Terrance D. *How to Design an Independent Power System.* Necedah, Wis.: Best Energy Systems for Tomorrow, 1981.

Planning Research Corporation. *Solar Photovoltaics Applications Seminar.* McLean, Va.: Planning Research Corporation, 1980.

Praul, Cynthia Grace; Marcus, William B.; and Dawson, Mary H. *Delivering Energy Services: New Challenges for California's Electric Utilities.* Sacramento: California Energy Commission, 1982.

Prepared for the California Energy Commission by Resource Dynamics Corporation. *Cogeneration Handbook.* Sacramento, Calif.: California Energy Commission, 1982.

Report to the Congress by the Comptroller General: Hydropower—An Energy Source Whose Time Has Come Again. Washington, D.C.: General Accounting Office, 1980.

Resource Planning Associates. *The PG and E Cogeneration Financial Analysis Program.* San Francisco: Pacific Gas and Electric Company, 1981.

Schneider, Daniel J., and Damstrom, Emory K. *The Schneider Hydroengine.* Justin, Tex.: Schneider Engine Company, 1980.

Schweppe, Fred C.; Tabors, Richard D.; and Kirtley, James L. *Homeostatic Control: The Utility/Customer Marketplace for Electric Power.* Cambridge: MIT Energy Laboratory, 1981. Order no. MIT-EL 81-033.

Selfridge, John, ed., *Kansas Wind Energy Handbook.* Topeka: Kansas Energy Office, 1981.

Sencenbaugh, James. *Sencenbaugh Wind Electric Design Manual.* Mt. View, Calif.: Sencenbaugh Electric Company, 1981.

Shafer, L., and Agostinelli, A. *Centrifugal Pumps as Hydraulic Turbines for the Small Hydropower Market.* Taneytown, Md.: Worthington Group, McGraw-Edison Company, 1981.

Solarex Corporation. *Making and Using Electricity from the Sun.* Blue Ridge Summit, Pa.: TAB Books, 1979.

Solomon, Lawrence. *Breaking Up Ontario Hydro's Monopoly.* Toronto: Energy Probe, 1982.

Staff of the Governor's Energy Council. *Pennsylvania Renewable Energy Resource Assessment.* Harrisburg: Governor's Energy Council, 1982.

Stern, Philip S. *Wasted Water Pressure and Potential Energy Generation: A Feasibility Study of the Hydroelectric Potential in Part of the Domestic Water System of Boulder, Colorado.* Boulder: Colorado Project/TIP, 1980. Available from Colorado Project/TIP, Inc., P.O. Box 731, Boulder, CO 80306.

Thaxter Lipex Stevens Broder & Micoleau. *Key Provisions of a Sample Purchase Power Agreement: A Guide for Hydro Developers.* Washington, D.C.: The National Alliance for Hydroelectric Energy, 1982.

Tudor Engineering Company. *Simplified Methodology for Economic Screening of Potential Low-Head Small-Capacity Hydroelectric Sites.* Palo Alto: Electric Power Research Institute, 1981.

U.S. Department of Energy. *The Energy Exchange Project.* Washington, D.C.: U.S. Department of Energy, 1982. Order no. DOE/CS/20468.

The Utility Role in Cogeneration and Small Power Production, Hearings Before the Subcommittee on Energy Conservation and Power of the Committee on Energy and Commerce, House of Representatives, Ninety-Seventh Congress, First Session April 27 and June 3, 1981. Washington, D.C.: U.S. Government Printing Office, 1981. Serial no. 97-56. Available from Committee on Energy and Commerce Documents, Room B-334, Rayburn House Office Building, Washington, DC 20515.

VanKuiken, J. C.; Buehring, W. A.; Huber, C. C.; and Hub, K. A. *Reliability, Energy and Cost Effects of Wind-Powered Generation Integrated with a Conventional Generating System.* Argonne, Ill.: Argonne National Laboratory, 1980. Order no. ANL/AA-17.

Wegley, H. L.; Ramsdell, J. V.; Orgill, M. M.; and Drake. R. L. *A Siting Handbook for Small Wind Energy Conversion Systems.* Richland, Wash.: Battelle Pacific Northwest Laboratories, 1980. Order no. PNL2521rev1.

Whisnant, R. A.; Morrison, C. B.; Staffa, N. G.; and Alberts, R. D. *Application Analysis and Photovoltaic System Conceptual Design for Service/Commercial/Institutional and Industrial Sectors: Final Report.* Vols. 1 and 2. Albuquerque: Sandia National Laboratories, 1979. Order no. SAND79-7020.

Wilcox, Delos. *Municipal Franchises.* New York: McGraw-Hill Book Co., vol. 1, 1910, and vol. 2, 1911.

Wilraker, V. F.; Eichler, C. H.; Hayes, T. P.; and Matthews, M. M. *Photovoltaic Utility/Customer Interface Study.* Albuquerque: Sandia National Laboratories, 1980. Order no. SAND80-7061.

Wind Energy: Investing in Our Energy Future. Sacramento: California Energy Commission, 1981.

Yamayee, Zia A., and Peschon, John. *System Integration Issues of Residential Solar Photovoltaic Systems.* Washington, D.C.: U.S. Department of Energy, 1980. Order no. DOE/RA/29349-01.

Index

*Page numbers in **boldface** type indicate tables.*